Small Gas Engine Repair

About the Author

Paul Dempsey is a master mechanic and the author of more than 20 technical books, including *How to Repair Briggs & Stratton Engines* and *Troubleshooting and Repairing Diesel Engines,* both in their Fourth Editions and both published by McGraw-Hill.

Small Gas Engine Repair

third edition

Paul Dempsey

New York Chicago San Francisco Lisbon London Madrid
Mexico City Milan New Delhi San Juan Seoul
Singapore Sydney Toronto

The McGraw-Hill Companies

Library of Congress Cataloging-in-Publication Data

Dempsey, Paul.
Small gas engine repair / Paul Dempsey.—3rd ed.
p. cm.
ISBN 978-0-07-149667-4 (alk. paper)
1. Small gasoline engines—Maintenance and repair. I. Title.
TJ790.D448 2008
621.43′40288—dc22 2008009828

1 2 3 4 5 6 7 8 9 0 DOC/DOC 0 1 4 3 2 1 0 9 8

ISBN 978-0-07-149667-4
MHID 0-07-149667-X

Sponsoring Editor: Larry S. Hager
Production Supervisor: Pamela A. Pelton
Editing Supervisor: Stephen M. Smith
Project Manager: Virginia Howe, Lone Wolf Enterprises, Ltd.
Copy Editor: Mick Spillane
Proofreader: Jacquie Wallace, Lone Wolf Enterprises, Ltd.
Art Director, Cover: Jeff Weeks
Composition: Lone Wolf Enterprises, Ltd.

Printed and bound by RR Donnelley.

This book is printed on acid-free paper.

Contents

Preface

I was re-reading Voltaire's *Candide* the other day and, as always, was struck by the ending. After experiencing all the horror that the 18th century was capable of—war, state- and church-sponsored torture, conscription, slavery—Candide and his battered crew conclude that "We must cultivate our garden." In other words, we cannot reform the world: our job is to tend to those tasks, however humble, that are within our power to accomplish.

The book you are holding in your hands is not about gardening, it is about maintaining the machines upon which our gardens, lawns, and so much else depend. But I think Voltaire would have approved.

There's a tremendous satisfaction in making what was a inert hunk of metal come alive. And, in the process, you can save hundreds or even thousands of dollars. The way we discard mowers, edgers, trimmers and other types of expensive engine-driven equipment is a national disgrace. According to Briggs & Stratton, the life of a walk-behind mower averages about 200 hours. At that point, the owner, frustrated because the machine won't start, rushes out and buys a new one. It would be just as easy and a whole lot cheaper to change the spark plug or clean the carburetor.

I don't have data on the life expectancy of riding mowers and garden tractors, but suspect that it's not much better. Drive by any working-class neighborhood and you can see inventories of lawn equipment rusting in the back yards. And all for the lack of a belt or a spark plug.

Small engines, in my experience at least, rarely wear out in the classic sense. Sometimes they fail early, because of owner negligence or abuse, but the overwhelming majority of malfunctions involve the accessories. Carburetors, starters, and ignition systems go out while the engine still has years of service left in it.

Anyone armed with patience and a few hand tools can make these repairs, once they know what's wrong. Engines don't talk or, more exactly, they have only two things to say—"You brilliant technician, I'm fixed!" or "No, stupid, that wasn't the problem." The trick is to break through this communication barrier.

Skill at diagnostics separates real mechanics from parts changers. Throwing parts at a problem eventually solves it, but the exercise gets expensive. Real mechanics, like good poker players, bet on probabilities.

How does one learn to do this? Part of the skill at diagnosis comes about from experience. No book, video, or school can teach the little tricks and the subliminal knowledge that comes about from spending years on the shop floor. But a book can explain how the technology works, and that's 90 percent of the battle. Take diaphragm carburetors, for example. These devices are unique to small engines and as alien to automotive carburetors as anything in Area 51. Without understanding how these little gems work, your chances of fixing one are about the same as winning the Texas lottery. In so far as this book has any lasting value, it is because it explains the often arcane technology of small engines.

But there's more to diagnostics than experience and theory. One also needs a systematic approach, a kind of opts manual to guide the troubleshooting process. Chapter 2 describes how to respond, in what I hope is an intelligent way, to the various ways in which engines fail. Subsequent chapters go into detail about troubleshooting ignition, fuel, starting, and charging systems.

The emphasis is on repair of Briggs, Tecumseh, Kohler and two-stroke edger and trimmer engines, with brief forays into Honda, Onan, and Wisconsin. Many of the procedures apply to other engines as well. Once you come to terms with, say, a Briggs & Stratton, a Kawasaki has few surprises.

Material on some of the longer-lived vintage engines is presented, but the focus is on current technology. About half of the text has been rewritten and updated for this edition of the book.

You will also find information here about buying parts on the Internet, fabricating special tools, and other ways to get engines back into service in the most economical way possible.

Paul Dempsey

1

Basics

This chapter is for readers new to small engines. Experienced mechanics can skip the next few pages, but everyone should read the section on safety at the end of the chapter. Small engine repair is not without its hazards.

Four-cycle engines

Construction

Figure 1-1 illustrates the internal components that make up a four-cycle engine. The example shown has a horizontal crankshaft, side valves, and splash lubrication. Crankshafts can be oriented horizontally or vertically. Horizontal-crank engines are used whenever the driven element is also horizontal. Vertical-crank engines, developed initially for direct-drive rotary mowers, also find application in garden tractors, tillers, and power washers.

As shown in the drawing, side-valve engines locate their valves alongside the cylinder bore. These engines remain in production, but overhead valve (ohv) and overhead cam engines are fast replacing them. The overhead, or I-head, configuration places the valves facedown directly over the piston. This arrangement makes for a compact, clean-burning combustion chamber, but adds complexity in the form of pushrods and rocker arms (Fig. 1-2).

Mounting the camshaft in the head simplifies the hardware. In 1995, Kohler introduced its 18-hp overhead cam (ohc) engine, which went on to earn the New Product Award from the National Society of Professional

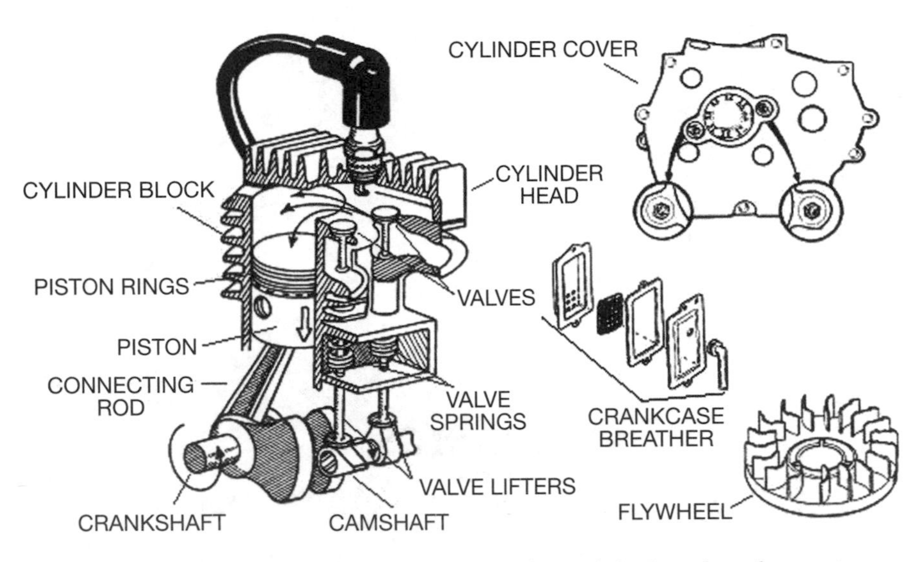

FIGURE 1-1. *Internal parts of a horizontal-crankshaft, side-valve engine.* Tecumseh Products Co.

FIGURE 1-2. *Pushrods and rocker arms actuate the valves on ohv engines.* Subaru Robin

Engineers. A previous winner was the Boeing 777. Honda, Kawasaki, and Subaru Robin quickly followed suit with their own ohc engines (Fig. 1-3).

Side-valve engines employ splash lubrication. Horizontal-shaft models have a dipper on the end of the connecting rod that splashes oil about the crankcase to lubricate the bearings and cylinder bore. Vertical-shaft engines use a camshaft-driven slinger—a kind of paddle wheel—to the same effect (Fig. 1-4).

Splash is generally adequate for side-valve engines that have their moving parts confined within the crankcase. Tecumseh, to its credit, supplements splash with a small, piston-type pump on its vertical-shaft models. The pump services the upper crankshaft and camshaft bearings that might otherwise starve for oil. Certain Briggs flathead engines incorporate a pump whose sole function is to circulate oil through the replaceable filter.

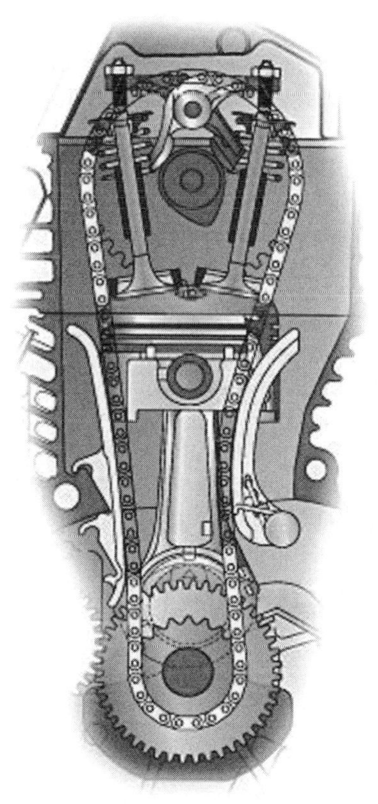

FIGURE 1-3. *Quality ohc engines use a chain, rather than a toothed belt, for the cam drive.* Subaru Robin

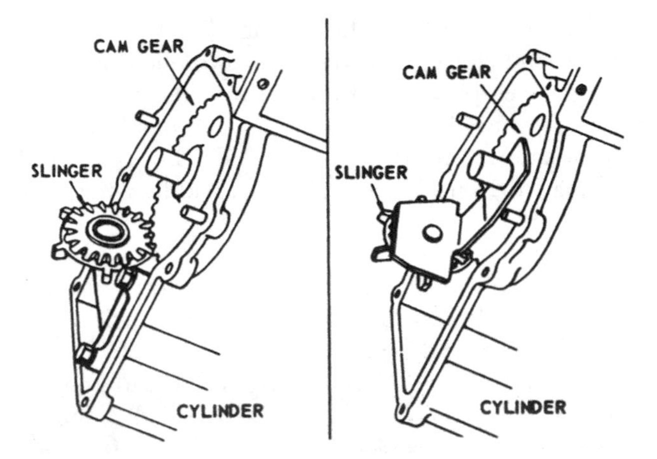

FIGURE 1-4. *Cam-driven oil slinger used on Briggs & Stratton vertical-shaft engines.*

Valves mounted in the cylinder head require some provision for lubrication. Most manufacturers supply oil to the valves with a pump, such as the one shown in Figure 1-5. Honda is an exception: Its ohc GC and GS models use the timing belt as a conveyor to move oil from the crankcase to the camshaft and valves.

Two main bearings support the crankshaft. The typical four-cycle runs its crank on aluminum block metal, which works well enough so long as the oil remains clean. Better-quality engines employ replaceable brass or Teflon bushings. Top-of-the-line models are fitted with ball bearings that run in deeply grooved races to support radial and thrust loads.

One of the more problematic features of small engines is the use of aluminum connecting rods that run directly against the crankshaft. Aluminum fatigues rapidly (which is why this material is not used for automotive rods) and is by no means the ideal bearing metal. Only a handful of Japanese engines provide proper bearings in the form of babbit-lined inserts.

Pistons normally carry three rings: one to distribute oil around the cylinder bore and two to seal compression.

An aluminum or cast-iron flywheel has magnets in its rim that energize the ignition system and, when fitted, an alternator. Fan blades on the flywheel generate airflow that, in conjunction with the sheet-metal shrouding, provides cooling.

FIGURE 1-5. *An Eaton-type oil pump of the kind favored by many small-engine manufacturers.*

Operation

Figure 1-6 illustrates piston and valve motion for the four-stroke-cycle. The sequence begins with the piston moving downward from top dead center (tdc), or the upper limit of its travel. Air and fuel enter past the open intake valve (A). The exhaust valve remains closed during this and the subsequent stroke.

The intake valve closes as the piston rounds bottom dead center (bdc) and begins its climb toward tdc. With both valves closed, the mixture undergoes compression as the piston rises (B). How much it is compressed is expressed as the ratio of the cylinder volume at the bottom of the stroke divided by the volume at tdc. Compression ratios range from about 6:1 for side-valve engines to slightly more than 8:1 for the newer ohv models. All things equal, the higher compression ratio, the greater the power output.

Once the air-fuel mixture is compressed, the spark plug fires to initiate the expansion stroke (C). The exhaust valve opens near bdc and remains open for the duration of the exhaust stroke (D). The valve then closes to set the stage for the next intake event.

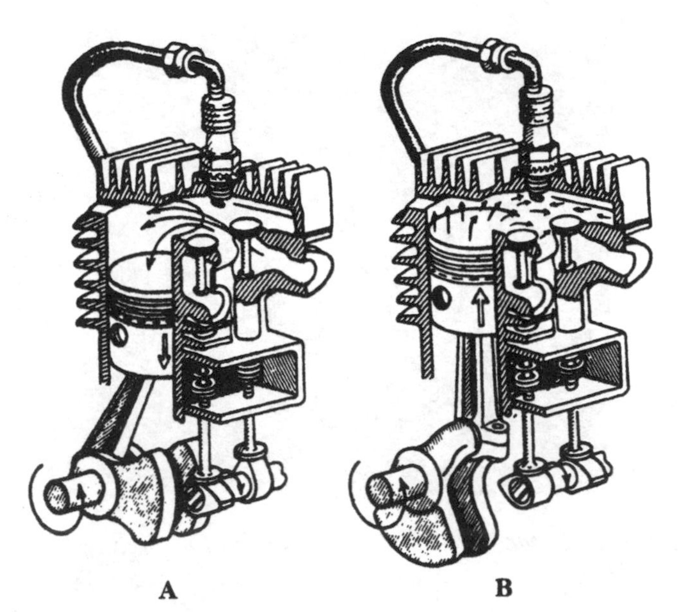

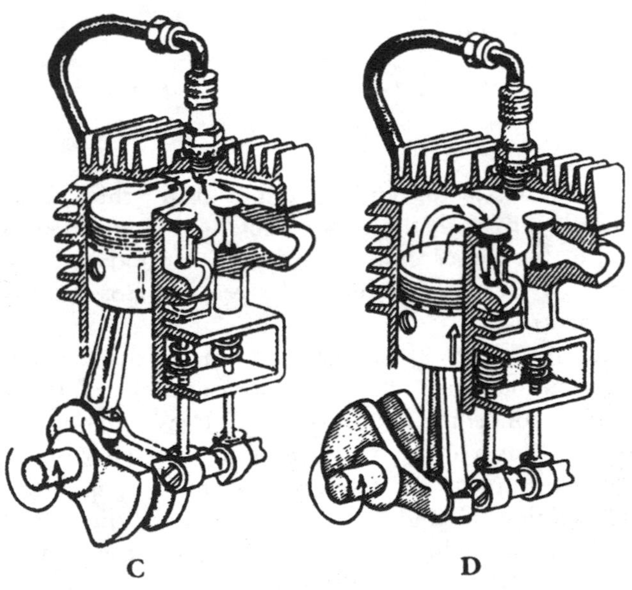

FIGURE 1-6. *The four-stroke-cycle requires two full crankshaft revolutions, or 720° of crankshaft movement, to complete. Events occur in this sequence: intake (A), compression (B), power or expansion (C), and exhaust (D).*

Two-cycle engines

Construction

Two-stroke-cycle engines cost little to manufacture and deliver 30 to 60 percent more power than equivalent four-strokes. Without oil in the crankcase, these engines run in any position, even inverted. Consequently, two-stroke engines furnish power for most portable tools, snowmobiles, small watercraft, and lightweight motorcycles.

The engine shown in Figure 1-7 is about as simple as internal combustion gets. It feeds and exhausts through ports that open into the cylinder bore. The cylinder head is integral with the block. A single piston ring generates the necessary compression. Edger, string trimmer, and other lightweight engines further simplify things by cantilevering their crankshafts off a single main bearing (Fig. 1-8).

Operation

The piston is the central player. In addition to its primary function as a converter of heat energy into mechanical motion, the piston acts as shuttle valve, opening and closing the cylinder ports. It also functions as a pump, drawing an air and fuel mixture into the crankcase and discharging it into the combustion chamber.

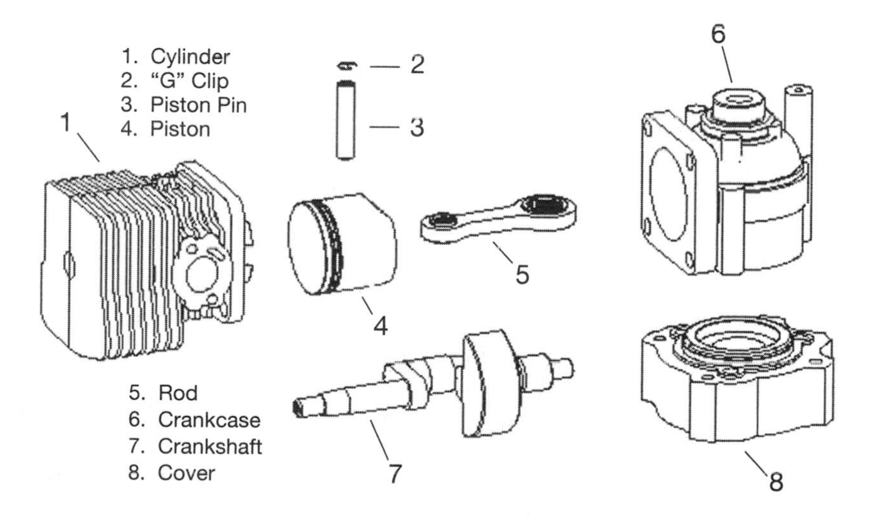

FIGURE 1-7. *A two-cycle Tecumseh engine, less carburetor, ignition system, flywheel, and shrouding. Ball and needle bearings permit lean 50:1 fuel/oil mixes.*

FIGURE 1-8. *Many two-cycle engines, including some of the better commercial models, cantilever their crankshafts off a single main bearing.*
Robert Shelby

The upper left drawing of Figure 1-9 shows the piston near bdc. Exhaust gases from the previous cycle flow over the piston crown and out to the atmosphere through the exhaust port. Simultaneously, the port labeled "intake" (most people would call it a transfer port) is open to connect the area above the piston with the crankcase, which has been pressurized by the falling piston. Air and fuel move through this port into the cylinder bore and, in the process, drive out, or scavenge, most of the remaining exhaust gases.

In the upper-right drawing, the piston climbs to compress the fuel charge ahead of it. The crankcase, previously evacuated and now undergoing an increase in volume, experiences a pressure drop. The reed valve responds to this pressure drop by opening to admit a fresh charge of fuel and air into the crankcase.

As the piston approaches tdc, the spark plug fires. The lower drawing shows the piston moving downward under the force of combustion gases. The reed valve closes and the falling piston pressurizes the crankcase. At this point, the crankshaft has made one full revolution and the operating cycle is complete.

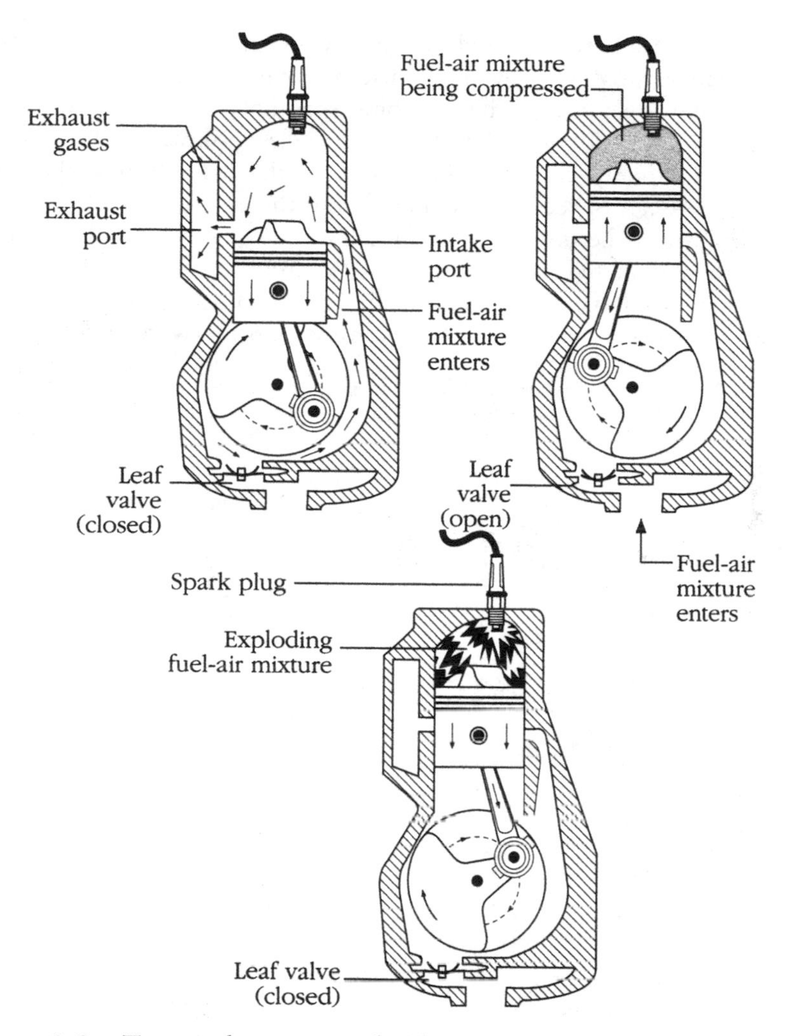

FIGURE 1-9. *Two-stroke or two-cycle (the terms are used interchangeably) engines complete the operating cycle in a single revolution of the crankshaft. In theory, a two-cycle engine should produce twice the power of an equivalent four-cycle, but the actual advantage is less.*

All two-cycle engines incorporate some form of crankcase valve to admit fuel and air when the piston is near the top of its stroke and to seal crankcase pressure as the piston falls. The engine illustrated employs a pressure-actuated reed valve—not unlike the reeds in a musical instrument—for this purpose. Another approach is to use the piston itself as the valve (Fig. 1-10). This "third-port" arrangement remains popular, although it does not provide the flat torque curve associated with reeds. High-revving engines are sometimes fitted with a rotary valve. A cutaway in the valve face opens the crankcase to the carburetor as the piston rises during the compression phase of the stroke.

A mechanic should remember that a two-cycle crankcase is a pressure vessel operating under a compression ratio of about 1.6:1. A leak at the crankshaft bearing seals or reed valve denies fuel to the engine.

Scavenging is the process of scrubbing exhaust gases from the cylinder. The example depicted back in Figure 1-9 is cross-scavenged, which means that the transfer and exhaust ports are directly opposite each other. A deflector on the piston crown diverts the incoming charge upward and away from

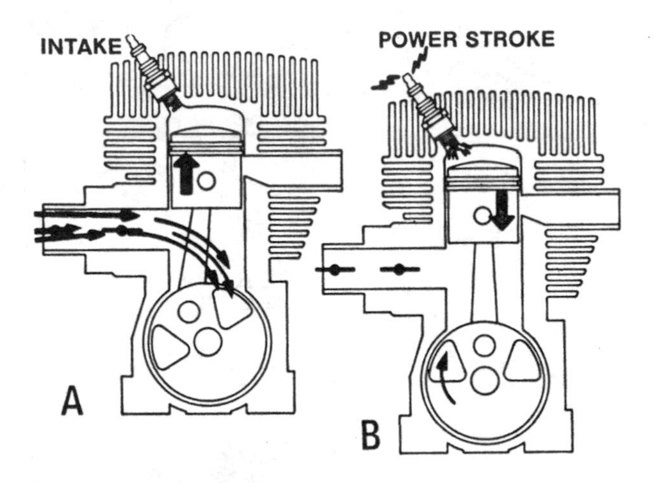

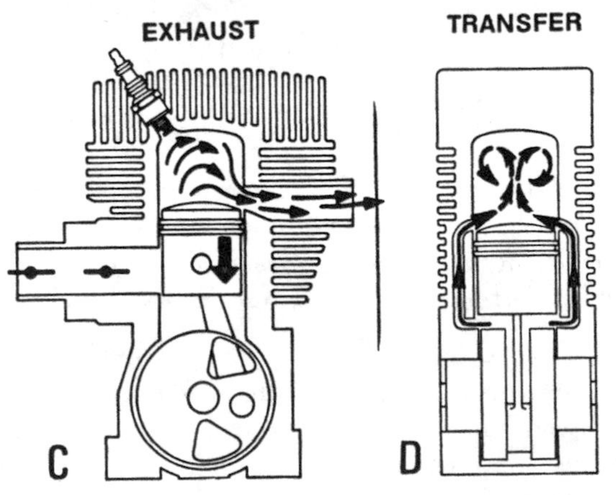

FIGURE 1-10. *Third-port engines remain popular for scooters and mopeds, although reed valves offer better control over the carburetion.* Walbro Engines

the open exhaust port. However, the deflector can only do so much: some exhaust gases remain in the cylinder and a large fraction of the fuel charge "short circuits" out the exhaust port. The wasted fuel dirties the exhaust and inert gases that elude scavenging contaminate the fresh charge. This latter phenomenon is responsible for "four-stroking" at idle, when scavenging is at its worst. The engine skips one or two beats and then ignites the accumulated fuel charge with a loud pop.

Loop scavenging, shown in Figure 1-11, represents a major improvement over cross-flow scavenging. Transfer ports are deployed radially around the cylinder circumference with angled exit ramps. These ramps direct the charge upward toward the domed cylinder head. The fuel streams converge and form a vortex that drives the exhaust gases out ahead of it. The miniature cyclone tends to remain in the chamber so that less fuel escapes out the exhaust port.

But scavenging is very imperfect. About a third of the fuel is wasted at wide throttle angles and as much as 70 percent at idle, when the velocity of the incoming charge is low. Much of the blame for the abysmal air quality of Asiatic cities can be attributed to two-stroke mini-motorcycles and tricycles.

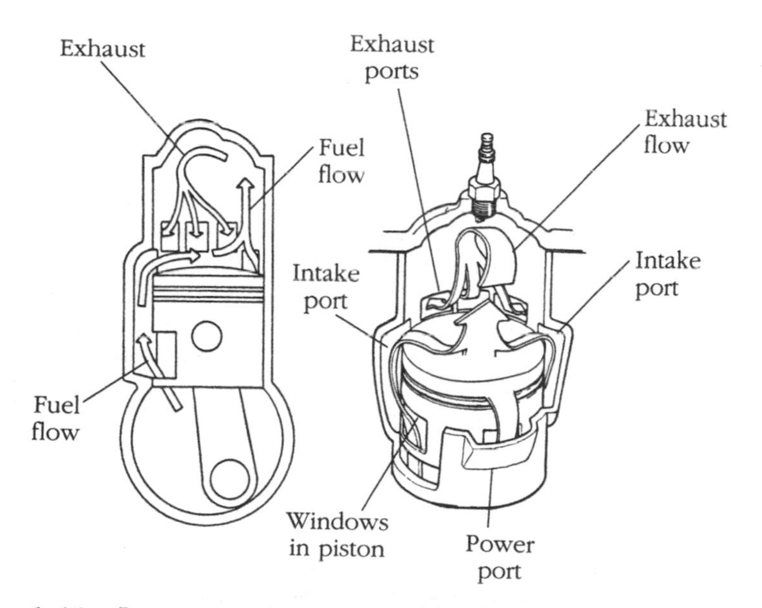

FIGURE 1-11. *Loop scavenging, pioneered by the German motorcycle manufacturer DKW and copied by everyone, was a major breakthrough in two-cycle design.*

The high level of exhaust emissions have led some manufacturers to abandon the two-stroke engine. Briggs & Stratton stopped making them 20 years ago. Briggs, Honda and Echo all produce mini-four strokes for weed trimmers and other portable tools. Other manufacturers, unwilling to give up the power and cost advantage of these engines, see direct injection (DI) as the solution.

DI means that fuel is injected late into the cylinder as the piston climbs during the compression phase. Since the exhaust port is closed, no fuel escapes combustion. This technology slashes hydrocarbon emissions (i.e., unburnt gasoline and oil) by as much as 90 percent and carbon monoxide by 70 percent. According to one study, two-stroke motorcycles with conventional carburetors averaged about 30 kilometers per liter of fuel.[1] When the same motorcycles were adapted to DI, fuel economy increased to 40.1 km/L.

This technology is expensive and, at present, applications are limited to a few upscale European and Taiwanese motor scooters and to Bajaj autorickshaws—three-wheeled commercial vehicles popular in India, Southeast Asia, and parts of Africa. But, at present, DI represents the best hope for the future of two-stroke engines.

Hybrid

Stihl, the Austrian firm best known for its chainsaws, has developed a hybrid engine that combines two-stroke crankcase induction with four-cycle operation (Fig. 1-12). The 4-Mix is, as far as I know, the first production four-cycle engine that uses the piston to generate a supercharge effect by pressurizing the crankcase.

The dry crankcase permits the engine to be operated at any angle, which is an important feature for handheld tools. Unlike most two-strokes, it easily meets EPA Phase 2, California Air Resources Board (CARB) Level 3, and current EURO emissions standards. The supercharge effect, plus efficient four-stroke scavenging, produces 17 percent more torque than an equivalent two-stroke. Power is up 5 percent and fuel economy 30 percent. While rotational speeds are high—peak power occurs at 10,000 rpm—the total-loss oiling system, combined with Teutonic quality, should make for a long-lived engine.

---

[1]"Direct Injection as a Retrofit Strategy for Reducing Emissions from 2-Stroke Cycle Engines in Asia," Dr. Brian Wilson, Dept. of Mechanical Engineering, Colorado State University, Fort Collins, Co, SAE 80523-1374.

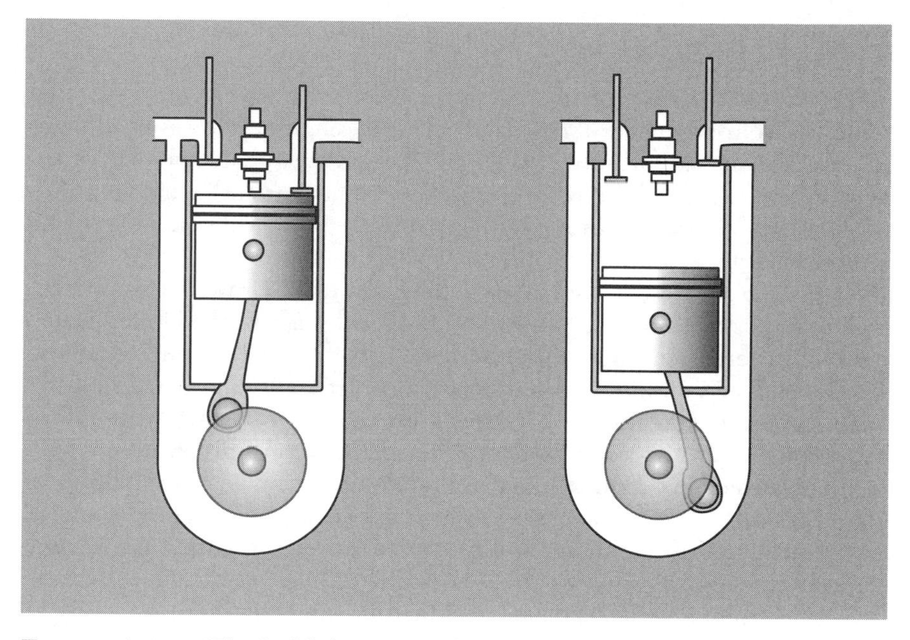

FIGURE 1-12. *The Stihl 4-Mix combines two-stroke crankcase scavenging with four-stroke operation. As shown on the right hand drawing, the exhaust valve opens as the piston rises on the exhaust stroke. At the same time, air and fuel enter the crankcase. In the left-hand drawing, the piston has rounded tdc on the intake stroke. The falling piston pressurizes the crankcase, forcing the air-fuel mixture past the open intake valve and into the combustion chamber.*

Displacement

Displacement—the cylinder volume swept by the piston—is the fundamental measure of engine potential in the way that square footage is to houses, tonnage is to ships, or caliber is to firearms.

The formula for calculating displacement is:

$$\text{bore} \times \text{bore} \times \text{number of cylinders} \times \text{stroke} \times .7858 = \text{displacement}$$

When bore and stroke are expressed in inches, the formula gives displacement in cubic inches (cid). The Briggs Sprint 96900 has a 2.56-in. bore and a 1.75-in. stroke:

$$2.56 \text{ in.} \times 2.56 \text{ in.} \times 1 \text{ cylinder} \times 1.75 \text{ in.} \times .7858 = 9.01 \text{ cid}$$

Power, torque, and RPM

What follows is a short excursion into mathematics, which might seem far removed from the practical details of repairing engines. However, the utility of these simple formulas should not be underestimated. In 1903, the Wright brothers used the same elementary equations and whatever information they could find in magazines to build the world's first successful aircraft engine.

Engine output has two components: torque and horsepower. Torque is the instantaneous twisting force applied on the crankshaft. In the United States, we express torque in units of pound-feet: 1 lb-ft. = 1 lb acting on a lever 1 ft. long. The rest of the world uses newton-meters: 1 N-m = 0.725 lb-ft.

Horsepower is the measure of work done over time. James Watt coined the terms in 1782 as a way of describing the utility of his steam engines. Watt observed that a mine pony, tethered to a capstan, lifted 550 lb of coal 1 ft. every second or 33,000 lb in 1 minute. A 1-horsepower engine would accomplish the same amount of work over the same time period. Expressed metrically 1 hp = 745 kW (kilowatt).

Torque, rpm, and horsepower have the following relationship:

$$(\text{Torque} \times \text{rpm})/5252 = \text{horsepower}$$

$$\text{Torque} = \text{displacement} \times 4\text{pi} \times \text{bmep}$$

The latter term, bmep (brake mean effective pressure), is the average pressure applied to the piston during the expansion stroke.

Horsepower expresses the ability to function under steady load, as when mowing a well-tended lawn, pumping water, or generating a constant amount of electric power. A jogger generates about 0.1 hp on flat pavement. Torque reveals itself as the ability to cope with sudden loads, as when a mower encounters a thick clump of Johnson grass or a mechanic heaves on a stubborn bolt.

Figure 1-13 graphs horsepower and torque against rpm for a Briggs Raptor kart engine. The Raptor develops maximum torque at between 3000 and 3500 rpm, when cylinder filling is at its most efficient. Maximum horsepower comes on line 1000 rpm later, which reflects the relationship between increased horsepower and engine speed. But beyond 4500 rpm, internal friction—most of it generated by the piston rings—costs power. As a racing engine, the Raptor is redlined at 6500 rpm, a speed that would be lethal for the standard product. Utility engines are governed to 3600 rpm or less, which is well below the power peak.

Horsepower and torque are always something less than advertised. American and foreign engine makers determine horsepower in accordance with

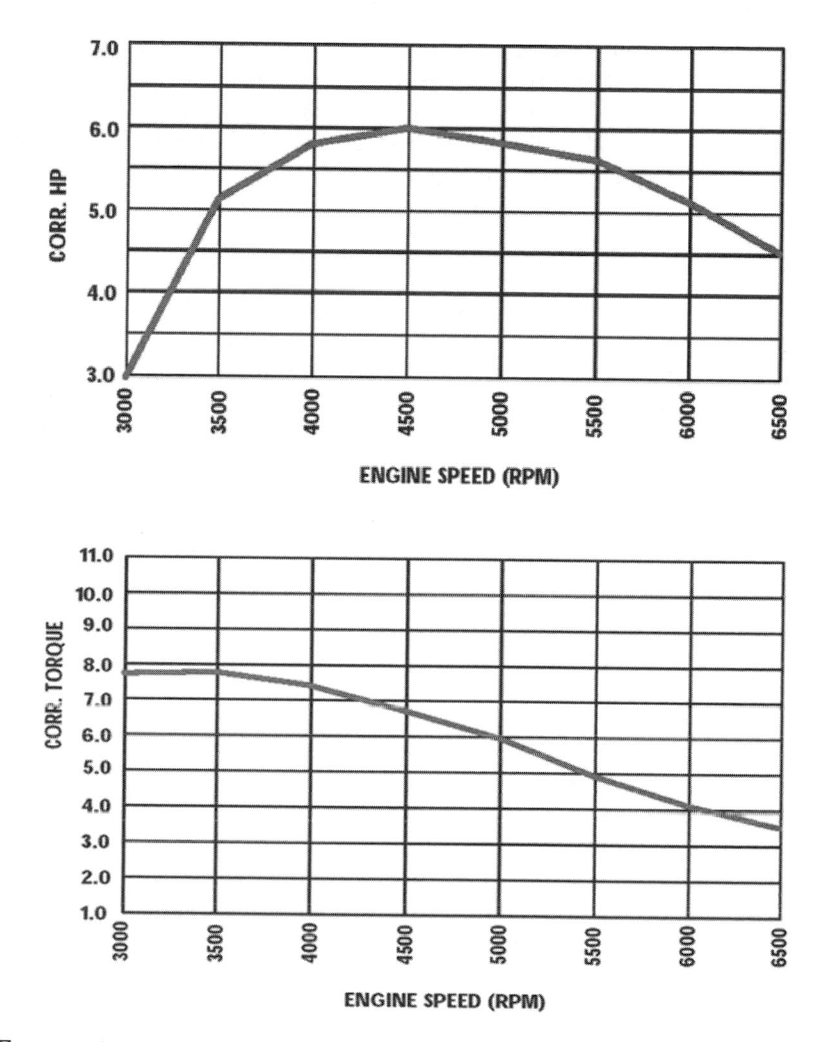

FIGURE 1-13. *Horsepower (A) and torque (B) curves for the Briggs Raptor.*

the SAE J1940 protocol. A sample engine is run on a dynamometer at a calculated altitude of 100 m (328 ft.) and at 25° C (77° F). Valves, carburetors, and ignition systems are tuned to laboratory standards and the engine is stripped of power-robbing accessories.

Production engines tuned to the same precision and after break-in and disassembly for carbon removal can be expected to deliver about 85 percent of advertised power. The applications manual for in-house Briggs' engineers suggests that 80 percent is a more realistic figure.

Even without special tuning, power drops off 3 percent for each 1000 ft. of altitude above sea level and 1 percent for each 10° F (5.6° C) above 60° F (15.6° C). However, the greatest inhibitor comes about because of limits on rpm. No engine survives long at the speed needed to develop full power. Many applications put additional constraints on rotational speed. For example, safety considerations limit rotary-mower blade tips to 19,000 ft./min. When coupled to a 24-in. blade, the maximum permissible engine speed must be governed to 3025 rpm.

While published horsepower and torque figures give a rough indication of performance, the real test comes about in the field. The engine should run smoothly under normal loads at no more than three-quarters throttle. An engine that baulks, hunkers down, and gasps for breath is underpowered for the application. Of course, the more power, the slower the engine can turn and, all things equal, the longer it will live. One-lung oil-field engines, ticking over at a few hundred rpm, run for decades with only routine oil and spark-plug changes.

Lubricating oils

Four cycle

Four-cycle engines use motor oils that are graded by viscosity and performance characteristics. Viscosity, or weight, refers to the oil's "pourability," with the higher numbers representing thicker oil. A 30-weight oil is more viscous than 10- or 20-weight oil.

Oil loses viscosity as temperature rises. Thus, oil that functions well at 100° F thickens and makes starting difficult at subzero temperatures. To overcome this difficulty, refiners developed multiviscosity oils, such as 5W-40 or 10W-40. The "W" stands for "winter," so 10W-40 behaves like 10-weight oil during cold-weather starts and gives the protection of 40-weight at high temperatures.

There is some argument about the use of multigrade oils in hot weather. Several manufacturers, including Briggs & Stratton and Tecumseh, recommend straight 30-weight oil for summer operation. The viscosity extenders that give multigrade oils their flexibility can break down under high temperatures and shear loads. When this happens the oil has the consistency of water. On the other hand, Kohler and Honda suggest that 10W-30 is appropriate for summer use.

Viscosity recommendations apply to conventional, petroleum-based oils. Multi-grade synthetic oils contain few viscosity extenders and, according to

Amsoil and other manufacturers of these products, can be used in lieu of straight petroleum oils. The choice is up to the owner.

The American Petroleum Institute (API) grades motor oils by scuff resistance, anti-wear properties, resistance to oxidation, and other characteristics. "S" grades apply to spark-ignition engines, "C" to compression-ignition, or diesel, engines. SJ oils were introduced back in 1996, SL in 2001 and the current SM oils in 2004. Each grade supersedes the previous grade. In other words, if the owner's manual recommends SE, a grade that is no longer available, use SL or SM from a major refiner.

It also should be pointed out that engine manufacturers do not agree about what constitutes a full crankcase. On engines with vertical fill plugs, it is customary to bring the oil level up to the third thread from the top. Horizontal-crank Hondas and Honda clones with their fill plugs at an angle should be topped off. Factory-recommended oil-change intervals are just that, recommendations. Honda, for example, suggests changing the oil at 100 operating hours or six months. That, like the 6000-mile intervals recommended for new cars, appears to be stretching things a bit. Twenty-five hours, the same interval recommended for experimental aircraft engines, gives better protection. No engine has ever been worn out by frequent oil changes.

Warning: The U.S. Environmental Protection Agency classifies used motor oil as a toxic substance, known to cause skin cancers upon repeated exposure. Dispose crankcase oil at a recycling center.

Two-cycle

When an engine starves for lubrication, the rod bearing usually goes first, followed by the main bearings. Early engines ran their aluminum (or bronze in the case of Jacobsen) rods directly against the crankpin and supported their crankshafts on brass bushings. Bearings of this type require copious amounts of oil. Sixteen parts of gasoline to one part oil was the standard mix and this standard still applies, regardless of the type of oil used.

In modern engines the bushings have been replaced with ball, roller, or needle bearings. These bearings, which make rolling contact with their journals, need almost no lubrication. Fifty parts fuel to one part API TC oil is generally adequate, although Tecumseh and a few other manufacturers insist on richer mixtures for some models. Table 1-1 converts fuel-oil ratios to liquid measures.

API TC lubricants are formulated for air-cooled two-cycle engines of up to 500-cc displacement. These synthetic or petroleum-based oils also meet Japanese (JACO FC) and European (ISO-L-EGO) standards. Lubricants labeled NMMC TC-W3, while intended for outboard motors, conform to API TC standards and are safe to use in air-cooled engines.

TABLE 1-1. Two-cycle fuel/oil mix

	U.S.		METRIC	
Ratio	**Gasoline**	**Oil to be added**	**Gasoline**	**Oil to be added**
24:1	1 gal	5.3 oz	4 L	167 mL
	2 gal	11.0 oz	8 L	333 mL
32:1	1 gal	4.0 oz	4 L	125 mL
	2 gal	8.0 oz	8 L	250 mL
40:1	1 gal	3.2 oz	4 L	100 mL
	2 gal	6.4 oz	8 L	200 mL
50:1	1 gal	2.5 oz	4 L	80 mL
	2 gal	5.0 oz	8 L	160 mL
100:1	1 gal	1.3 oz	4 L	40 mL
	2 gal	2.6 oz	8 L	80 mL

Buying parts

To purchase parts you will need the engine's model, type and serial numbers, and, in some cases, the build date. This information will be stamped on the block or on a tag affixed to the cooling shroud (Fig. 1-14).

Factory dealers are the best sources of parts, although the markup is high and sometimes arbitrary. A Google search will provide parts breakdowns for most small engines, together with vendors whose prices are generally lower than you would pay at an authorized dealer. Some of these parts are OEM (original equipment manufacturer); others are sourced from the aftermarket. For example, Yamakoyo, a Chinese manufacturer with a Japanese-sounding name, makes copies of various Honda engines. Many of the parts interchange with the Honda originals, and cost about half as much as Honda charges. While Yamakoyo parts appear to be of reasonable quality, other aftermarket components are little more than junk. In the junk category are starter motors with skimpy insulation, pleated-paper oil filters that are open at the edges like the pages of a book, and ignition modules that fail within hours of startup.

Buying an engine

While manufactures like Kohler, Subaru Robin, and Cummins Onan concentrate on the upper end of the market, high-volume producers such

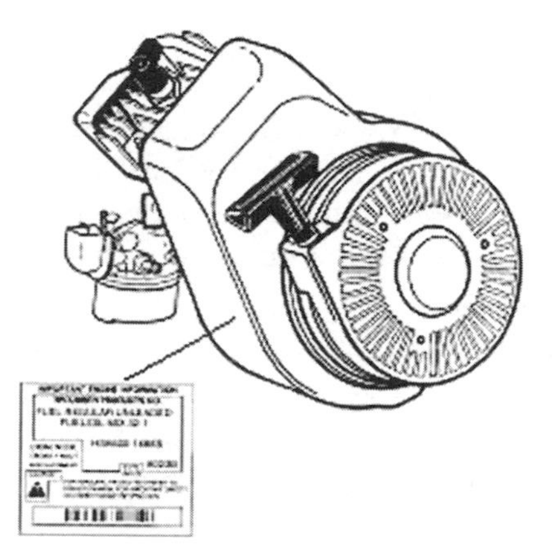

FIGURE 1-14. *Each manufacturer has its own ID protocols. Information typically includes engine model, type number, customer code number, build date, and emissions certification.* Tecumseh Products Co.

as Briggs & Stratton, Tecumseh, Honda, and Echo build for every pocketbook. Consequently, you can buy a good Briggs, one that is worthy of the 100-year-old history of the company, or a mechanical fruit fly with an estimated life of 200 hours. The same goes for products from the other mass marketers.

So what does one look for in an engine?

- Cast-iron cylinder liners. Throwaways run their pistons directly on the aluminum block.
- Overhead valves. While the ohv configuration says nothing directly about quality—some have aluminum bores—these engines are more fuel efficient and less polluting than their side-valve predecessors. And many have other desirable features such as oil filters, efficient air cleaners, and quiet mufflers.
- Centrifugal governors. Mechanical governors are more sensitive than air-vane types, which means less lugging and fewer shutdowns under sudden loads.
- Some form of pressurized lubrication. With the exception of Honda and Kawasaki overhead cam units, all valve-in-head engines incorporate an oil pump. However, be aware. Some Briggs engines include a pump that has no function other than to supply oil to the filter.
- An automotive-type carburetor, mounted away from the fuel tank with a conical float bowl. Diaphragm carburetors represent, in the writer's opinion, a tradeoff between maneuverability and reliability. That is,

these carburetors permit trimmers, chainsaws, and other portable tools to be operated at any angle off the vertical. But diaphragms need frequent replacement and have no place on lawnmower or utility engines.
- A two-piece air filter that incorporates a sponge-type precleaner and a pleated-paper element on four-cycle engines. Two-stroke engines, because of the oil fog that hovers around the carburetor throat, cannot use the more efficient paper filters.

The most convenient way to purchase an engine is from a dealer, who should be able to help you decide on the best choice for your application. Liquidators are another source. Volume buyers, such as Murray or Toro, sometimes overestimate their engine requirements. Engines not sold by the end of the season end up in the hands of liquidators. This is big business. Kansas City Small Engines has a turnover of four million dollars a year and carries something like 40,000 units in inventory. These engines sell for well below dealer list. Granger and Northern Tool are also sources of discounted engines, but stocks are limited to the most popular units. "Factory seconds," engines without warranties, can also be purchased on the Internet. As far as I know, these engines are not seconds in the classic definition of the term. That is, they passed final inspection, but were subsequently damaged during shipment. Expect to find minor damage such as dented shrouds and broken spark plugs.

Safety

The chief hazard associated with small-engine repair is spilled gasoline. Work in a well-ventilated area, away from possible ignition sources. Wipe up spills immediately and allow ample time for the residue to evaporate.

Never refuel or open a fuel line on a hot engine. Winter blends of gasoline ignite at temperatures slightly above the boiling point of water. Your local gasoline distributor can verify that statement.

Asbestos was used at least until the 1980s for gaskets and clutch facings. Some of these parts are, no doubt, still on dealer shelves. And what the Chinese use is anyone's guess. One way to deal with this material—although I cannot guarantee its absolute safety—is to grease gaskets before removal. Scrape the gasket off with a razor blade and dispose of the fragments in a sealed container. Do not attack suspect gaskets with a wire brush or wheel.

Use conventional solvents, such as kerosene or Gunk. The latter product, available from auto parts stores, rinses off with water. While not perfectly safe—California authorities have identified Gunk as a carcinogen—Gunk seems to pose less of a hazard than most other commercial solvents.

Keep hands and finger clear of V-belts and other moving parts. Do not work on the business end of lawnmowers, tillers, chippers, and other rotating equipment without first defeating the ignition. In some cases, merely disconnecting the spark-plug lead is not sufficient: the wire "remembers" where it has been and floats back into proximity with the spark plug. Wedge the connector into the cylinder fins or, better; ground it with an alligator clip.

Warnings, Cautions, and Notes follow military practice. That is, a Warning means risk of personal injury, a Caution means risk of equipment damage, and a Note is a comment about some point of interest.

2

Troubleshooting

Some mechanics can look at an engine, run a few tests, and tell you what's wrong with it. More often than not, they are right. Others, like gamblers down on their luck, throw progressively more expensive parts at the problem. This chapter attempts to take some of the uncertainty out of troubleshooting.

Tools and supplies

Essentials for troubleshooting are:

- A fresh supply of fuel.
- Paper towels or lint-free shop rags.
- Varsol, kerosene or one of the biodegradable detergents sold at auto-parts stores.
- Wynn's Carburetor Cleaner or an equivalent product.
- At least one spare spark plug for the engine in question.
- Basic mechanic's tools in both English and metric sizes. You will also need a gauge for gapping spark plugs (Fig. 2-1), an ignition tester (Fig. 2-2), a flywheel puller, and when working on rotary mowers, a hand pump.

Preliminaries

Diagnostics is an exercise in information retrieval. Begin by learning as much as you can about the nature of the problem and events that led up to it. What exactly is the complaint: Hard starting? Sudden shutdowns?

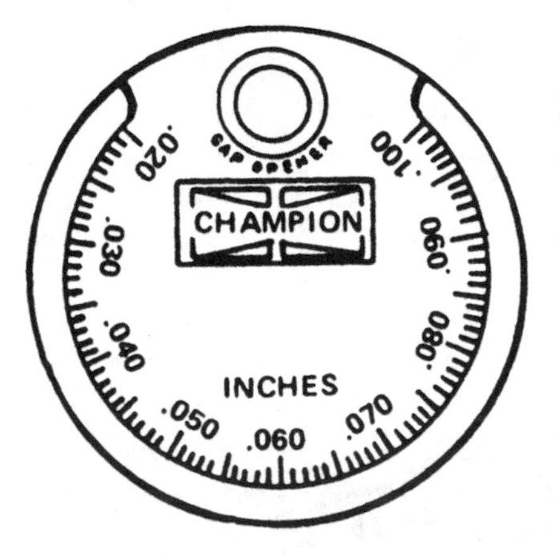

FIGURE 2-1. *A ramp-type gauge, such as the one shown here, quickly and accurately sets spark-plug electrode gaps. If you use a conventional flat-blade feeler gauge, bracket the readings. That is, when the specification calls for 0.030 in., adjust the gap so a 0.029-in. blade slips easily between the electrodes and a 0.031-in. blade generates noticeable drag.*

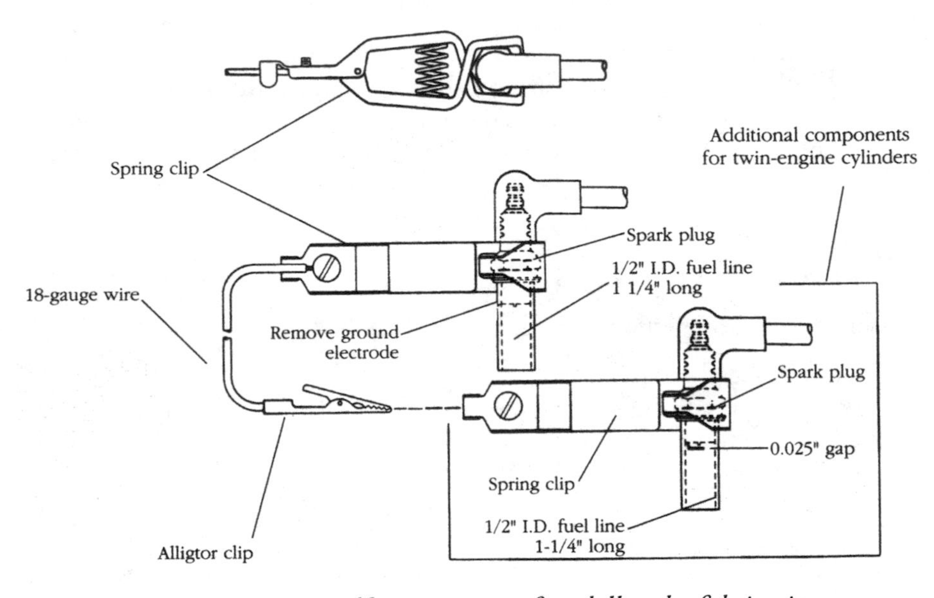

FIGURE 2-2. *Do-it-yourselfers can save a few dollars by fabricating an ignition tester from a spark plug with its side (ground) electrode removed, two alligator clips, a short length of wire, and a piece of fuel hose. The smaller alligator clip attaches to a cylinder-head fin; the hose shields the spark so it can be seen more easily. As it stands, this tester works on all engines, except Briggs with Magnetron ignitions. When testing these low-voltage systems, leave the side electrode intact and narrow the gap to about 0.15 in.* Kohler Co.

Lack of power? Excessive vibration? Did the malfunction occur suddenly or did it develop slowly, worsening over time? How long has it been since the engine was started? Were any repairs made just prior to the malfunction? If so, you can be almost certain that the mechanic did something wrong. What, if any, efforts were made to correct the problem?

Begin by checking the oil on four-cycle engines. Syrupy, black goop that feels gritty when rubbed between the fingers is a sure sign of trouble. You may want to remove the shroud and test for wear by pushing the flywheel from side to side. Normally, a crankshaft has about 0.002 in. radial play, or just enough movement to be perceptible. Greater movement, sometimes accompanied by audible clicks as the crank slaps against its bearings, means major repairs are in order. In extreme cases, the rim of the flywheel exhibits wear streaks from contact with the ignition-coil armature.

If there is any doubt about fuel quality, take the engine outdoors, drain the tank and refill with fresh fuel from a clean container. Gasoline has a shelf life of about six months. Stale gasoline, that is, gasoline that has an acrid smell and a brownish color, is responsible for most carburetor and valve problems.

Diagnostics will go more smoothly and with less cranking if sacrificial parts are changed early. These parts include:

- **Spark plugs** are the most frequent cause of starting difficulties. Unless the spark plug is out-of-the-box new, replace it with one of the correct heat range and type. Do not be misled by appearances: The plug may look clean and work perfectly outside of the cylinder, but fail to fire under compression. On rare occasions, even new spark plugs fail to function. Why this happens is unclear, but it probably has to do with the low cranking voltages developed by many small-engine ignition systems. You may want to test the spark plug in a running engine before using it as a diagnostic aid.
- **Fuel filter.** Use the correct factory part. "Universal" filters, intended for automobiles, clog quickly when used in low-pressure or gravity-fed small-engine systems.
- **Air filters.** Clean polyurethane foam filters in hot water and detergent. Dry and knead a tablespoon of motor oil into the element. Replace paper filter elements.
- If an electric starter is fitted, clean the battery terminals and verify battery condition as described in Chap. 6. Recharge if necessary. It should be noted that small-engine starters are undersized for the task and rapidly overheat. Give the motor several minutes to cool between 10-second bouts of cranking. Most mechanics prefer to use the rewind starter.

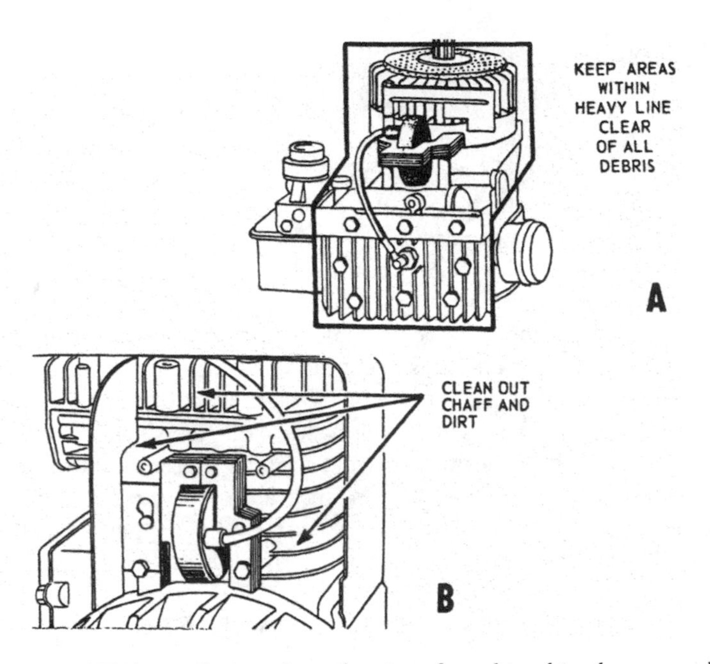

FIGURE 2-3. *While professional mechanics often skip this chore, good practice is to remove the shrouding and clean the cylinder fins whenever an engine comes into the shop.* Briggs & Stratton Corp.

Replacing these parts provides the opportunity to become acquainted with the equipment. Large, flakey rust blisters on the muffler and heavy corrosion on unpainted aluminum surfaces suggest that the machine was stored outside. Expect to find corrosion on electrical contacts, binding control cables, and rusted governor springs.

Heavy accumulation of oil on the cooling fins of four-cycle engines usually means that the owner overfilled the sump, although leaking gaskets and crankshaft seals cannot be ruled out. In any event, the shrouding should be removed and the fins cleaned (Fig. 2-3).

Loose blower housings and/or carburetor mounting bolts indicate excessive vibration, usually traceable to a bent crankshaft or rotary-mower blade.

Check the wiring to become familiar with safety features that could affect performance. Some engines have a solenoid-operated valve on the bottom of the carburetor that, should it malfunction, blocks fuel entry. Many four-cycles incorporate a sensor that denies ignition if the oil level is low or if the engine is tilted at extreme angles. A wire running out the crankcase will reveal the presence of such a sensor (Fig. 2-4). As described in the following

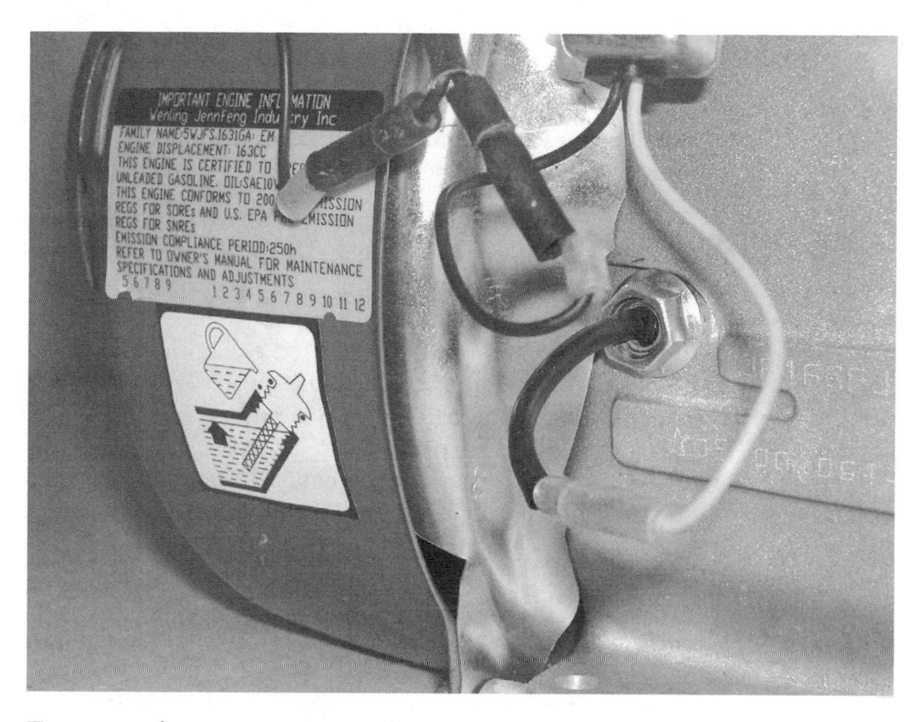

FIGURE 2-4. *A wire running from the crankcase means that the engine is equipped with an oil-level monitor that illuminates a warning lamp or, as is most often the case, shuts down the ignition.* Robert Shelby

chapter, garden tractors and riding mowers are fitted with safety interlocks that either open or ground the ignition circuit.

At this point, one should have a good idea of the general condition of the equipment and any special features that complicate diagnosis.

Engine binds or freezes during cranking

Disconnect and ground the spark-plug cable, and try to turn the flywheel by hand. Check the driven equipment for overly tight drive belts, misaligned shafts, frozen bearings, partially disengaged clutches, and any condition that generates drag.

The fault may be with the starter. Electric starters and associated circuitry should be tested as described in Chap. 6. Rewind starters may need to be centered over the flywheel hub by repositioning the cooling shroud.

If the engine itself is the source of the problem, major repairs are in order. See Chap. 7.

Engine cranks but does not start

In order to run, an engine must have spark, fuel, and at least 60 psi of compression. One or more of these prerequisites is lacking if a cold engine refuses to start after three or four pulls on the starter cord.

Spark test

Remove the recently replaced spark plug and, with the controls set on "Run," test spark output. Figure 2-5 shows a Briggs & Stratton PN 19051 spark tester that, unlike the homemade unit pictured earlier, is shielded to prevent accidental ignition of fuel spills. Two spark gaps are provided; the smaller gap is used with the Magnetron ignition modules that have been standard on Briggs engines since the early 1980s.

Crank the engine and watch for spark. The quality of the spark depends on the type of system: Magnetos and CDI systems produce thick, healthy sparks that blister paint. Briggs Magnetron systems require as much as 350 rpm to generate a spindly, reddish spark that is difficult to see in daylight.

You should see a steady shower of sparks—one per crankshaft revolution—as the flywheel is spun. No spark or an erratic spark means that the ignition system has failed.

Even if you have spark, it is good practice to check the condition of the flywheel key on rotary mowers. Striking a hard object with the blade can distort or shear the key. A sheared key takes out the ignition; a distorted key may permit the system to generate spark, but upsets timing enough to make starting difficult or impossible. Remove the cooling shroud, flywheel nut, rewind starter cup, and lockwasher. Verify that the slot in the flywheel hub aligns with the slot in the crankshaft. If the keyways do not align, lift the flywheel as described in the following chapter and inspect the key for damage.

Primer and choke

After the ignition, the next most likely culprit is the cold-starting system. Many carburetors use a primer pump to richen the mixture for starting (Fig. 2-6A). Remove the air cleaner and depress the rubber primer bulb three or four times. The pump should inject a stream of fuel into the carburetor bore. If it does not, turn to Chap. 4 for this and other fuel system repairs.

Other carburetors have a butterfly choke just aft of the air cleaner. The butterfly must close *fully* for cold starting (Fig. 2-6B). Most utility engines integrate the choke with the throttle, so that the choke closes when the throttle lever is past full open. Failure to close can usually be corrected by loosening the screw that secures the control cable to the engine (shown in the upper left of Fig. 2-6B) and moving the cable a fraction of an inch toward the choke.

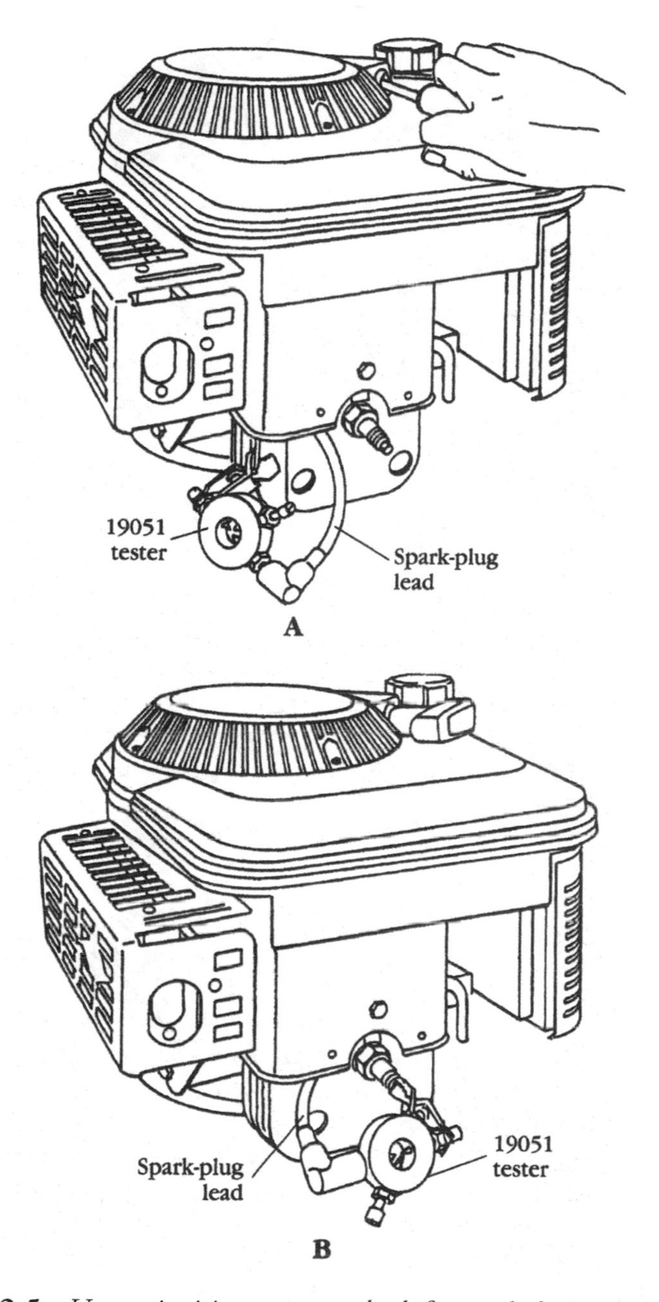

FIGURE 2-5. *Use an ignition tester to check for spark during cranking (A) and to detect voltage interruptions in a running engine (B). The Briggs PN 19053 tool illustrated eliminates a potential fire hazard by confining the spark behind a window.*

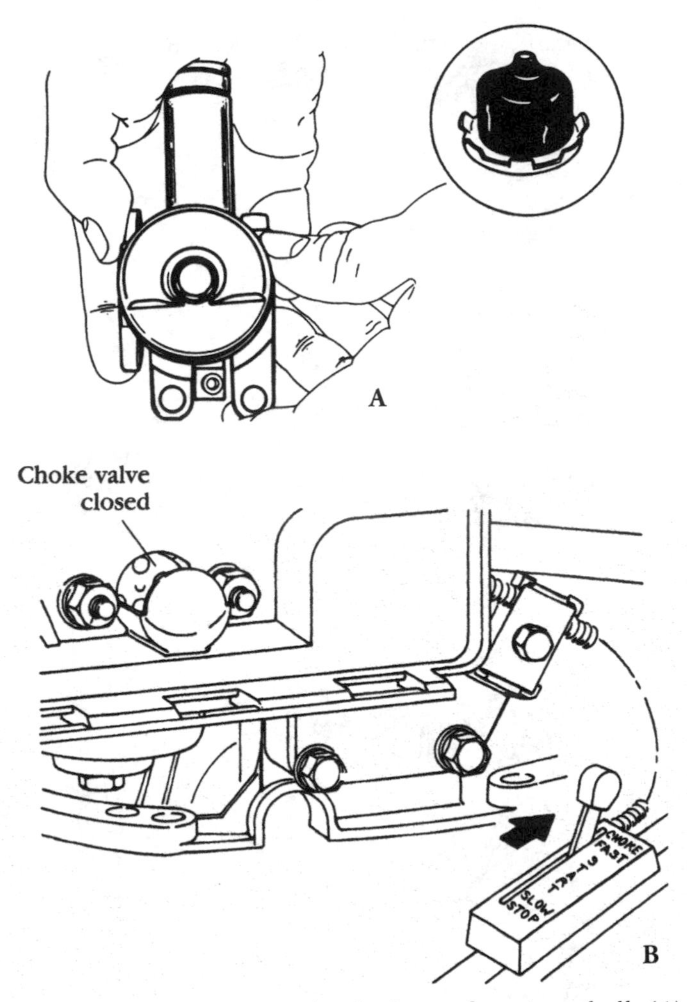

FIGURE 2-6. *Cold-start systems take the form of a primer bulb (A) or a choke valve (B). Depressing the bulb should flood the carburetor. The choke valve, or butterfly, must close fully for cold starting.* Tecumseh Products Co. (A) and Briggs & Stratton Corp. (B)

No or insufficient fuel

If the spark plug remains dry after a half-dozen choke-on starting attempts, take the machine outdoors, remove the air cleaner, and spray a small amount of Wynn's Carburetor Cleaner or an equivalent product into the carburetor bore. Replace the filter and crank.

Warning: Do start an engine without the air cleaner in place. The cleaner acts as a flame arrestor to confine backfires within the carburetor bore.

If the engine runs for a few seconds on carburetor cleaner and dies, you can be sure that the problem is fuel starvation. The fact that it runs at all indicates the presence of compression and spark.

Older carburetors have one or two adjustable jets that may have been tampered with. As described in Chap. 4, back out the adjustment screws one and a half to four turns from lightly seated and see if the engine will start.

Stand-alone carburetors receive fuel through a flexible hose. Most feed by gravity from the tank, but larger four-stroke engines often have a fuel pump, shutoff valve, and filter.

Working outdoors, disconnect the fuel line at the carburetor. The presence of fuel at this connection means that the stoppage is inside of the carburetor. If fuel does not reach the carburetor, work backwards toward the tank, disconnecting each fitting-pump output, pump input, filter output, filter input, and so on until you locate the problem.

Tank-mounted, suction-lift carburetors used on inexpensive Briggs engines tend to develop fuel stoppages as the result of hardened diaphragms or clogged pickup tubes. See Chap. 4 for details.

Fuel flooding

Flooding is easy to detect on a cold engine, since the spark plug will be damp and reek of gasoline. In extreme cases, fuel runs out of the carburetor air horn. But note that any engine floods if cranked long and hard enough, or if cranked when hot with the choke closed. Mechanics often flood engines while trying to fix them.

Clear the surplus fuel by removing the spark plug and blowing out the cylinder with compressed air. Or simply wait an hour or so for the fuel to evaporate. Install a dry spark plug, and with the choke and throttle full open, attempt to start the engine. We want to ingest as much air as possible. Flooded two-stroke engines are difficult to get back into operation because raw fuel puddles in the crankcase. It can be helpful to incorporate a spark gap into the ignition circuit. A Briggs spark tester, connected as shown in Figure 2-5 with the 0.166 gap in series with the ignition lead, boosts coil output by 13,000 V.

If flooding persists, the carburetor is at fault. Repair as described in Chap. 4.

Oil flooding

An oil-fouled spark plug means that crankcase oil has found its way into the combustion chamber. Any four-cycle engine can be made to oil flood if the mechanic cranks long and hard enough. However, the classic type of oil flooding occurs when an engine is tilted nose down, as when servicing a modern rotary mower without first emptying the crankcase. (Earlier mowers avoided this difficulty by mounting their engines with the spark plug aft.)

Oil flooding cures itself if the engine stands idle for a few days. A quicker solution is to blow down the cylinder with compressed air. The oil that remains can be cleared by spraying moderate amounts of carburetor cleaner into the spark-plug port. The starting drill requires a supply of clean spark plugs, which are replaced as they oil over. Eventually, the engine will come to life in a cloud of blue smoke.

Compression check

With the spark-plug terminal disconnected and grounded, crank the engine over a few times. Resistance on the cord should build and fall off as the piston rounds tdc under compression. Similarly, the sound made by the electric starter should change in pitch under compression. A "dead" cord or a steady hum from the starter motor suggests that compression is a problem. See Chap. 7.

Flywheel inertia

Engines with insufficient flywheel mass—a category that includes inexpensive mowers and other garden tools—impart a nasty feel to the starter cord. The cord bites back and attempts to retract as the motor is pulled through. The skimpy flywheels fitted to most rotary mowers require assistance from the blade. These machines will not start unless the blade is installed and bolted down securely.

Bite and pop back can also be caused by a failed compression release. Kawasaki, Briggs, Tecumseh, and most other utility engines employ automatic compression releases to reduce loads on the starter. These devices work by unseating the intake or exhaust valve during cranking. If the mechanism wears or if valve lash is excessive, the valve remains seated and starting becomes virtually impossible. See Chap. 7 for more information.

Long shots

At this point, we have pretty well covered all bases. If you have still not discovered why the engine refuses to run, give the job a rest. Removing yourself from the immediacy of the problem has a way of clarifying things.

An engine should—must—run if it has spark, fuel, and compression. But the real world has a way of confounding our generalizations. Two-stroke engines need five or six psi of crankcase compression to start. Some mechanics claim to be able to feel crankcase compression as they pull the engine through. The rest of us have to test crankcase integrity as described in the next chapter. Even so, it should be emphasized that lack of crankcase compression is a remote possibility, entertained only after the more likely suspects have been questioned and cleared.

Another long shot applies to overhead valve four-strokes that have been fueled with stale gasoline. Should one of the valves gum over and stick, cranking the engine bends the associated push rod. The engine will have compression and spark, but will fail to ingest fuel.

A similar problem has been reported for belt-driven Honda camshafts. A worn belt can permit the camshaft to jump time.

All bets are off if someone has gone into the engine since it ran last. A common mistake is to assemble four-cycle engines out of time. Another is to use the wrong parts. I remember as Briggs that kept the shop bus for days trying to discover why it had no spark. The complete ignition system was replaced to no avail. Finally, someone noticed that the flywheel magnets were positioned incorrectly relative to the ignition coil. Although the flywheel fit, it was the wrong part for the application.

Engine runs several minutes and quits

Connect a Briggs PN 19051 or an equivalent tool in series with the spark plug and start the engine. Watch the arcing in the window. If ignition failure is the problem, the flywheel will coast to a stop without generating spark.

Another possibility is fuel starvation, which will be indicated by a bone-white spark plug tip. This condition is best addressed by cleaning the carburetor, fuel pump, and replacing all diaphragms used in the system. A clogged fuel-cap vent can also shut the engine down after several minutes of operation.

Failure to idle

Failure to run at small throttle angles is usually a carburetor problem. If carburetor adjustments (described in Chap. 4) do cure the problem, disassemble the carburetor for cleaning with special attention to the low-speed circuit. A mal-adjusted governor can have the same effect.

Note: Do not try for a “Cadillac” idle. Modern four-cycle engines idle, if that is the correct term, at 1200–1400 rpm. A ragged idle, punctuated by loud pops, is normal for two-strokes.

Loss of power

First, determine whether the engine or the equipment it drives is at fault. Rotate driven equipment by hand to detect possible binds caused by failed bearings, misaligned shafts, dragging clutches, and overly tight v-belts.

Connect PN 19051 in series with the spark plug and run the engine under load to detect possible misfiring. If the ignition checks out okay, remove the spark plug. A lean mixture stains the tip of the plug bone white and may produce pop-backs and flat spots during acceleration. Richen the mixture as described in Chap. 4. Should the problem persist, clean the carburetor and look for air leaks between the carburetor and engine.

Note: If the engine requires choke to develop best power, you can be sure that it is not receiving sufficient fuel. Leaking seals on two-stroke engines can lean the mixture by ingesting air.

An overly rich mixture leaves fluffy black carbon deposits on the plug and, when pronounced, colors the exhaust with puffs of black smoke. Verify that the air filter is clean and that the choke opens fully. If possible, lean out the high-speed jet. The correct mixture produces brownish black deposits—the color of coffee with a dash of cream—on the spark plug tip.

Another possibility is that the governor spring, the spring that transfers force from the throttle lever to the carburetor throttle butterfly, has lost tension. When this happens, the engine will not run at rated rpm. Replace the spring with *the correct factory part.*

Warning: The wrong governor spring can cause the engine to over-speed and grenade.

Exhaust restrictions also cost power. Some applications require a spark arrestor in the form of a coarse-meshed wire screen upstream of the muffler. Periodically clean the screen with a soft wire brush.

Carbon also collects on two-stroke exhaust ports and mufflers. Steel mufflers can be cleaned in a bath of hot water and household lye. As for the ports, lower the piston below port height, and scrape off the deposits with a soft brass or copper tool, as shown in Figure 2-7. When finished, remove the

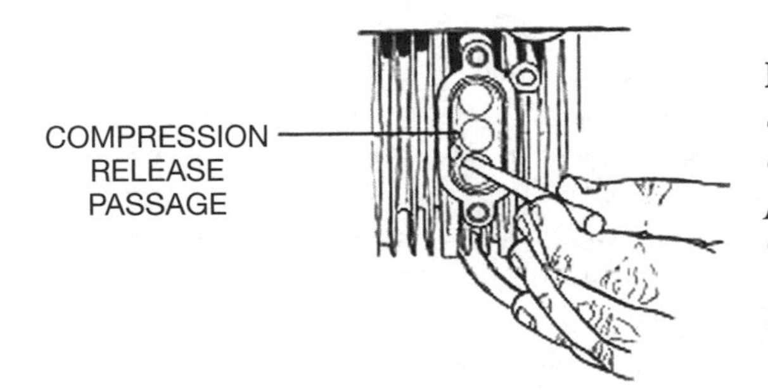

FIGURE 2-7. *Two-stroke exhaust ports collect carbon and should be periodically cleaned. Some Tecumseh engines incorporate a compression bleed port that may also carbon over.*

spark plug, ground the ignition, and spin the flywheel to clear the cylinder of loose carbon.

Excessive vibration

It is the nature of single-cylinder engines to vibrate. However, vibration that loosens bolts and generally makes things unpleasant is usually traceable to a bent crankshaft and/or rotary-mower blade.

When dealing with rotary mowers, empty the fuel tank to prevent spillage, remove the spark plug, ground the ignition, and tilt the machine up on its front wheels. As mentioned earlier, engines mounted head-forward oil flood, unless the crankcase is first pumped out.

Warning: Do not work on the underside of a mower, tiller, shredder, or other hazardous equipment without first removing the spark plug or grounding the ignition with an alligator clip. Do not trust the kill switch. Should the ignition function, any movement of the driven elements can start the engine.

Mark a point on the deck adjacent to a blade tip. Rotate the flywheel 180° and verify that the other blade tip aligns with the mark. If not, remove the blade and blade adapter. Place the blade on a flat surface. Bends or twists will be obvious. To determine if the crank is bent, focus on the bolt hole in the end of the crankshaft while a helper spins the engine over with the spark plug removed. Perceptible wobble means that a new crankshaft is in order. See Chap. 7 for further information.

Exhaust smoke

Black smoke means the engine is receiving more fuel than it can find oxygen to burn. Check that the choke opens fully, that the carburetor is properly adjusted, and that the air filter is clean.

Blue smoke is the sign of oil burning. Modern two-strokes produce no more than a wisp of smoke under acceleration when fueled with 40:1 or 50:1 premix. Worn piston rings are the most common cause of four-cycle oil burning. Leaking valve seals on overhead-valve engines generate smoke immediately after startup. Four-cycle crankcases operate under a slight vacuum generated by a check valve in the breather assembly. A malfunctioning check valve, loose dipstick, or air leaks destroy the vacuum and permit oil to migrate into the combustion chamber.

3

Ignition systems

Solid-state ignition systems need little by way of service other than routine spark-plug changes. Magneto and battery-and-coil systems do not have the same level of reliability.

Spark plug

Replace the spark plug every 100 hours of operation or at the first sign of hard starting. Most spark plugs fail as the result of carbon deposits that bleed ignition voltage to ground. When spark plugs must be changed frequently, verify that the plugs are the correct type and that the ignition system delivers consistent spark. Misfires foul spark plugs. Other causes of early spark-plug failure include overly rich fuel mixtures and excessive oil consumption.

Most mechanics do not attempt to clean spark plugs. However, a wire brush will remove carbon from the tip, which may be enough to get the engine started. Deeper deposits come off after several days' immersion in Permatex Carburetor & Parts Cleaner or an equivalent product. Do not sandblast spark plugs. Some abrasive invariably finds its way into the engine.

The interface between the spark-plug gasket and the cylinder head acts as a heat sink. Remove all traces of oil and grease from this area. Some mechanics like to spray the spark-plug threads with silicone to prevent sticking. Run the spark plug three full turns in by hand to prevent cross-threading. Torque specifications vary, but 210 lb/in. is appropriate for aluminum heads. The cast-iron heads on vintage engines require more torque, on the order of 300 in./lb.

Battered or carbon-clogged threads can be restored with an M14 × 1.25 metric tap. Heli-Coil® inserts make quick work of stripped threads, but the cost of the tooling makes this a job for an automotive machinist.

Flywheel

It is necessary to lift the flywheel to access the crankshaft key and under-flywheel ignition systems. Disconnect and ground the spark-plug lead and, on electric-start models, disconnect the positive (red) battery terminal. Remove the shrouding, starter motor, flywheel brake, and other components that block access to the flywheel.

A nut or, in the case of Briggs "heritage" models, the starter clutch secures the flywheel to the crankshaft stub. Figure 3-1 illustrates the two factory-

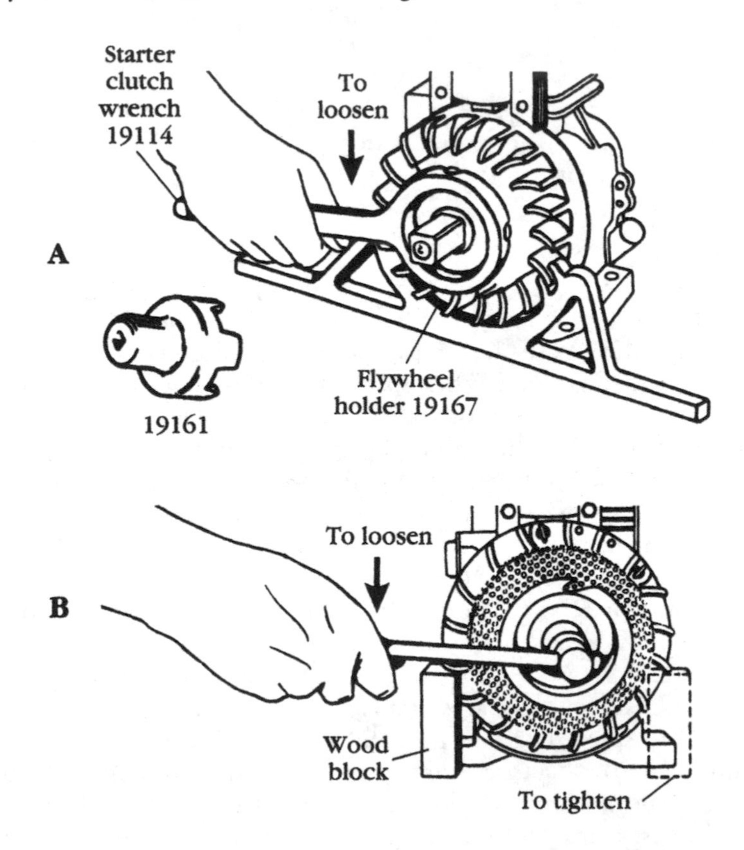

FIGURE 3-1. *A starter clutch (A) or hex nut (B) secures the flywheel to the crankshaft. The PN 19161 socket is the preferred clutch tool, since it can be used with a 1/2-in. drive torque wrench.*

supplied clutch wrenches. Clutches have been hammered on and off with a hardwood block.

Except for certain vintage horizontal-shaft engines, flywheel nuts have standard right-hand threads. When in doubt, trace the lay of the threads that extend past the flywheel nut. Flywheel-holding tools, such as the one illustrated in Figure 3-1, are of limited utility because flywheel diameters vary and many have plastic fans. A strap wrench works for all flywheels (Fig. 3-2). Rotary-mower crankshafts can be blocked from turning with a short piece of 2 × 4 between the blade and mower deck. Shop mechanics sidestep the problem with pneumatic impact wrenches.

Once the nut comes off, remove the rewind starter cup, noting how the cup indexes with the flywheel. Some engines have a Bellville-style lock-washer between the starter cup and flywheel. The convex side of the washer bears against the cup.

Utility-engine flywheels have provision for a puller in the form of two or three holes near the hub. For the most part manufacturers are considerate enough to thread the holes. Figure 3-3 illustrates one of several types of pullers.

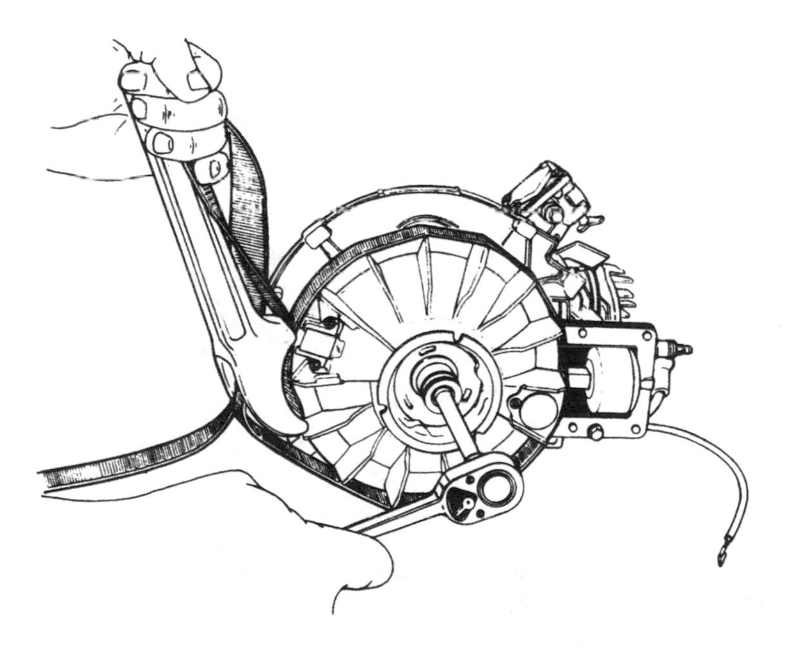

FIGURE 3-2. *Use a strap wrench to restrain the flywheel against retainer-nut torque. As mentioned in the text, a piece of 2 × 4 prevents rotary-mower blades from turning. When all else fails, a pipe wrench on the power takeoff end of the crankshaft makes a persuasive backup.*

FIGURE 3-3. *Proprietary flywheel pullers are available for many small engines, but an automotive harmonic balancer works better than most and adapts to two- and three-bolt hubs. Briggs flywheel hubs must be threaded for 5/16 × 18 bolts. Some European engines have a threaded counterbore in the hub that the puller makes up to. Bicycle shops sometimes can supply the tooling.* Robert Shelby

Caution: Do not attempt to pull a flywheel with a conventional gear puller that takes purchase on the flywheel rim. Force must be applied to the hub.

Other flywheels, including those for small two-strokes, must be shocked off. Tecumseh supplies three knockers—PN 670105 for right-handed 1/2-in. 20 threads, PN 670118 for the left-hand variant of the same thread, and PN 670169 for right-hand-thread 7/16-in. diameter shafts. Seat the knocker against the flywheel and back off three turns (Fig. 3-4). Insert a large screwdriver under the flywheel (clear of the ignition coil and other vulnerable components), and give the knocker a sharp rap with a hammer. Hit squarely and with sufficient force to compress the shaft taper and release the wheel. A glancing blow can break the cast-iron crankshaft.

Warning: Wear eye protection when hammering against steel.

A heavy brass bar or a brass hammer can substitute for a factory knocker. Protect the threads with the flywheel nut.

As shown in Figure 3-5, knocking off the flywheel can dislocate the crankshaft on two-stroke and other engines with anti-friction bearings. A blow

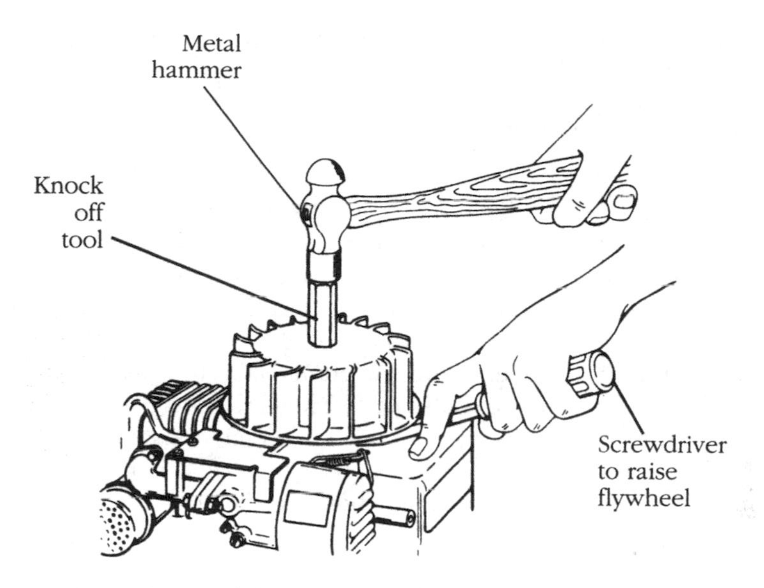

FIGURE 3-4. *Millions of flywheels have been successfully knocked off, but the procedure entails risk. Seat the Tecumseh-supplied knocker against the flywheel, back off a few turns, and hit squarely. A glancing blow can snap the crankshaft off.*

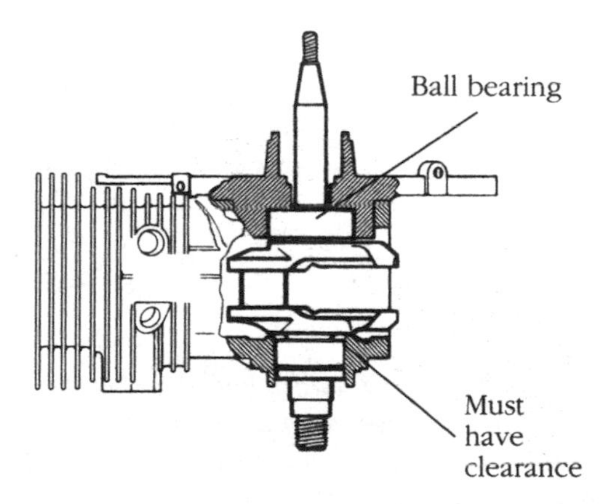

FIGURE 3-5. *Knocker-induced bearing dislocation can be corrected with a blow from a rawhide mallet on the pto end of the crankshaft. Engines affected, like the Tecumseh shown, have anti-friction bearings.*

with a rawhide mallet on the pto (power takeoff) end of the crankshaft restores end float.

Flywheel key and keyways

Carefully inspect the flywheel hub for cracks that nearly always radiate outward from the keyway (Fig. 3-6). Briggs uses soft aluminum keys to protect against flywheel damage; other manufacturers employ steel keys.

Warning: Always replace a cracked or otherwise damaged flywheel. Once a crack starts, it continues to grow until critical length is reached and the flywheel explodes.

Remove the flywheel key from the crankshaft stub, using side-cutting diagonal pliers if the key is stubborn. The key locates the magnets cast into the flywheel relative to the ignition coil. This relationship determines ignition timing for solid-state systems and synchronizes flux buildup with point openings for magnetos. A few thousandths of an inch of key distor-

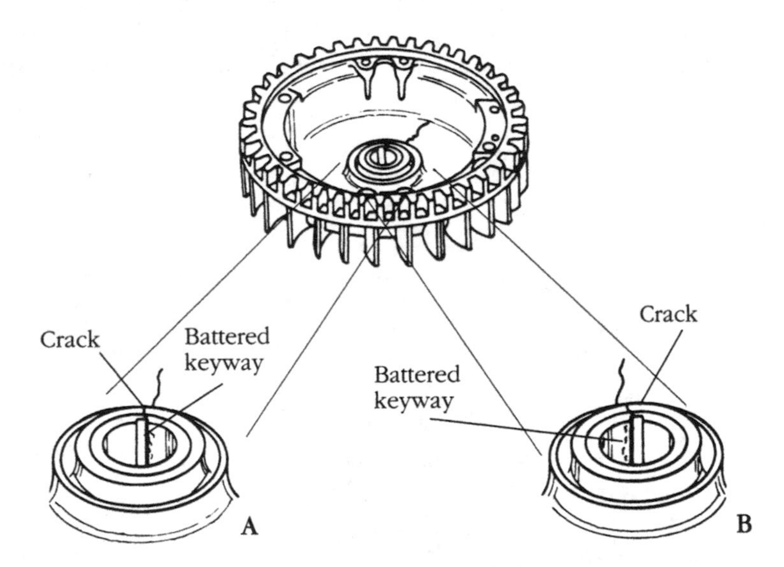

FIGURE 3-6. *A cracked flywheel is bad news, but there is some consolation in knowing how it happened. Assume that the crankshaft turns clockwise when seen from the flywheel. A crack on the leading edge of the keyway (A) means that the crankshaft over-sped the flywheel because of a loose hold-down nut. A crack on the trailing edge (B) suggests that the crankshaft stopped or slowed, allowing the flywheel to overtake it. Expect to find collateral damage, including a bent crankshaft and rotary-mower blade.*

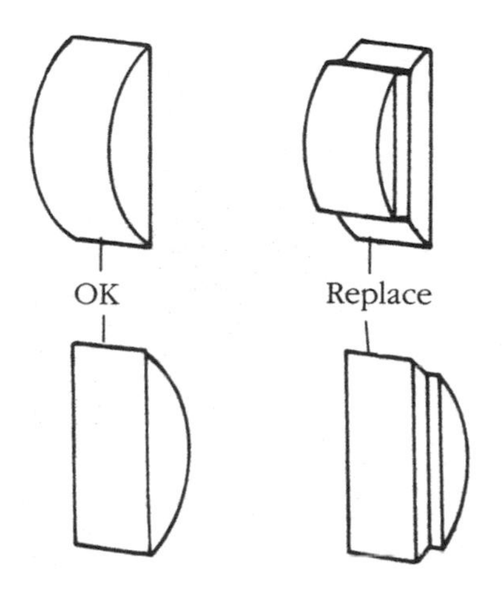

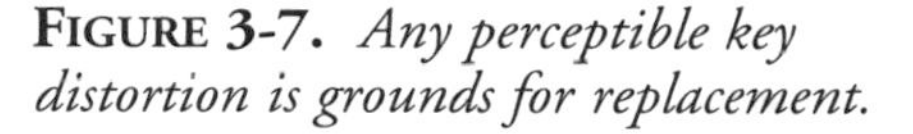

FIGURE 3-7. *Any perceptible key distortion is grounds for replacement.*

tion or slop between the key and keyways translates into a major error at the flywheel rim (Fig. 3-7). Timing errors can cause kickback as the rewind starter is pulled through, hard starting, and loss of power. A sheared key denies ignition.

Manufacturers recommend that the crankshaft and/or the flywheel be replaced when keyways exhibit perceptible wallow. No doubt, new parts represent the surest fix. But with careful assembly, the keyways can be aligned and the flywheel nut tightened to produce something close to original timing.

For the sake of completeness, it should be mentioned that flywheel magnets can weaken in service, although this type of failure rarely occurs with modern engines. A healthy magnet should attract a loosely held screwdriver through an air gap of 5/8 in. The smaller magnets used to energize CDI trigger coils need not be that powerful.

Spark test

Insert the key, mount the flywheel on the crankshaft, and run the nut down finger-tight. With the spark plug removed and an ignition tester connected to the high-tension lead, spin the wheel by hand. A vigorous spin should produce a spark.

Flywheel installation

Clean all traces of grease and oxidation from the tapers and remove any burrs. Install the starter cup, lockwasher (or Bellville washer with convex side up), and nut, lightly lubricated with 30-weight motor oil. Figure 3-8 shows how square and Woodruff keys should be installed.

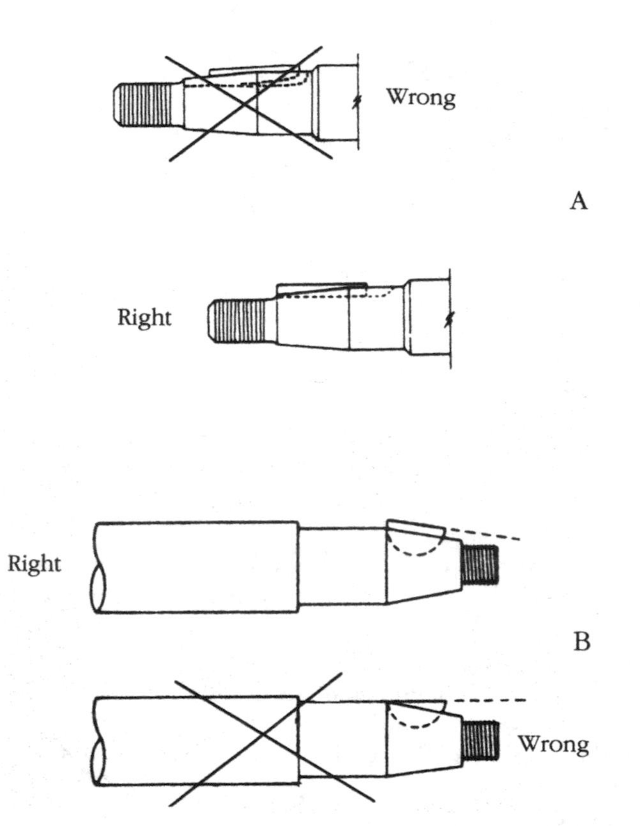

FIGURE 3-8. *Correct installation for Kohler (A) and OMC Woodruff keys (B). Briggs keys drop into place after the flywheel is mounted.*

Torque specifications vary with make and model, but, as a rule of thumb, tighten flywheel nuts as follows:

- Engines of less than 6 CID (cubic inch displacement)–40 lb/ft.
- 6 to 10 CID–55 to 60 lb/ft.
- 11 to 20 CID–85 to 90 lb/ft.

Basic CDI

Capacitive discharge ignition (CDI) has made conventional ignition systems, with their troublesome points and rpm-sensitive spark outputs, obsolete (Fig. 3-9).

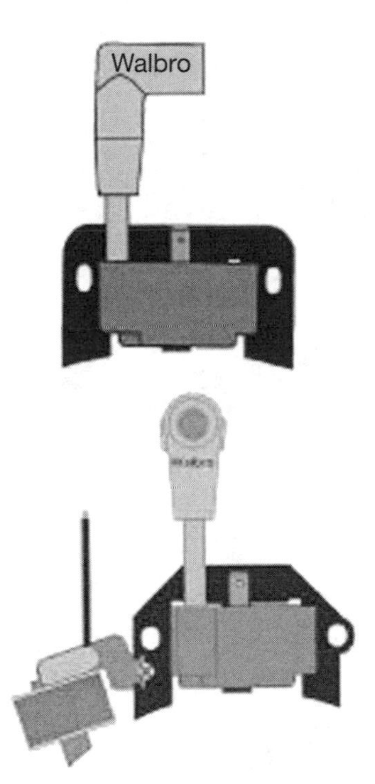

FIGURE 3-9. *Walbro digital capacitive discharge ignition module features customized timing (retarded spark to prevent kickback during cranking, advance to match engine requirements), idle stability, and precise governed speed control.*

Operation

Figure 3-10 illustrates the basic circuit. The flywheel magnet (1A) generates 200 VAC in the input coil (2). The rectifier (3) converts this AC output to DC for storage in the capacitor (4). The silicon-controlled rectifier (SCR at 7) remains non-conductive to block capacitor discharge.

About 180° of crankshaft rotation later, the flywheel magnet sweeps past the trigger coil (5) to generate a signal voltage across the resistor (6). This voltage causes the SCR to conduct. The stored charge on the capacitor (4) discharges through the primary side of the pulse transformer (8). Current flow in the primary side of the pulse transformer (actually an ignition coil) generates a 25,000 V potential in the secondary windings that goes to ground across the spark-plug electrodes.

Caution: Solid-state components are vulnerable to stray and reversed-polarity voltages.

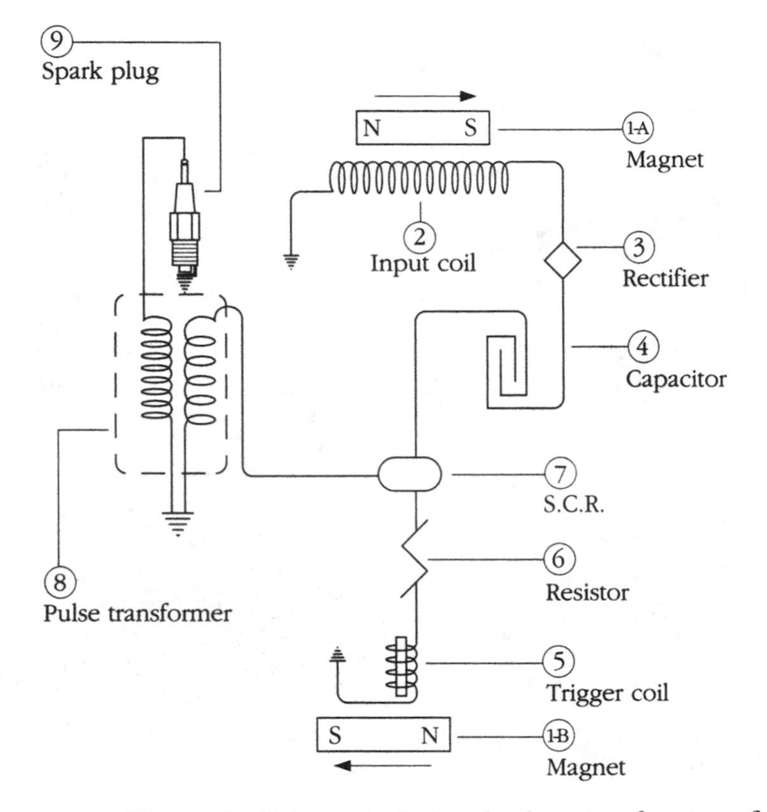

FIGURE 3-10. *Tecumseh CDI circuit. Induced voltage is a function of how rapidly the magnetic flux lines move across a conductor. At low engine speeds, the flywheel magnets must come into close proximity with the trigger coil to initiate ignition. As speeds increase, a weaker magnetic field suffices, and ignition occurs early, before the magnets align with the trigger coil. The system illustrated advances ignition timing in lock step with rpm. More sophisticated ignition modules shape the advance curve to better fit engine requirements.*

- When a battery is supplied, the negative (black) terminal goes to chassis ground. Reversing polarity is death on CDI and charging-system components, unless the system incorporates blocking diodes (Fig. 3-11).
- Do not operate the engine with the battery disconnected. Many applications use the battery as a voltage-limiting ballast resistor.

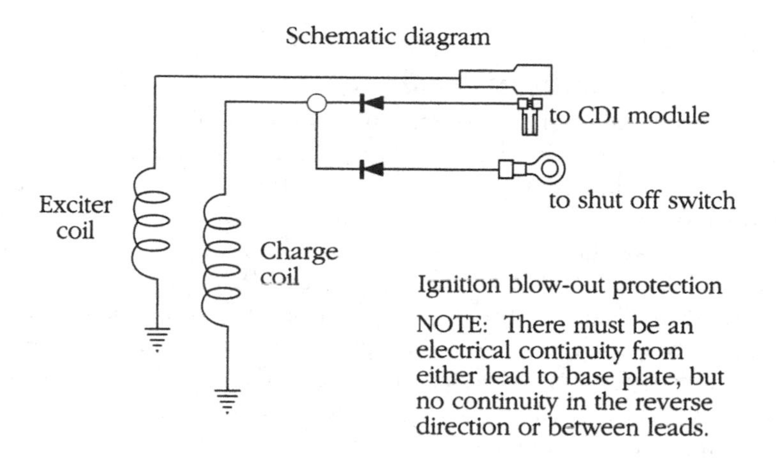

FIGURE 3-11. *Some CDIs employ diodes—the electronic equivalent of check valves—to protect against reversed polarity.*

- Do not disconnect or ground primary wiring when the engine is running or being cranked. Components that ground through their hold-down bolts must remain attached to the engine.
- Do not leave the spark-plug lead disconnected when cranking. Ground the ignition through a spark-gap tester. Open-circuit CDI voltages can puncture coil insulation.
- Finally, do not introduce stray voltages by welding on the engine or driven equipment.

Troubleshooting

- Replace the spark plug with a known good one of the same type as originally specified.
- Replace the flywheel key if distorted or sheared.
- Check for breaks in the wiring, worn insulation, loose harness connectors and oxidation on connectors and at engine/chassis grounding lugs. These ground connections should be tight and free of paint, grease, and rust. Silicon-based dielectric grease, available from auto-parts stores, protects connections from water intrusion.
- Verify that ignition interlocks function as described at the end of this chapter.
- Some dealers have equipment for testing CDIs, but the definitive test is to replace the suspect unit with a known good one.

Smart Spark

The Kohler Smart Spark CDI advances spark timing with engine speed. As indicated earlier, all solid-state systems have some built-in advance capability, since an increase in flywheel velocity induces voltage earlier in the trigger coil. The Kohler system quantifies the amount of advance to more accurately track rpm.

Figure 3-12 sketches the main features of a Smart Spark CDI. Flywheel magnets induce an AC voltage on the input coil (L1). Part of the coil output passes through diode (D1) for rectification and to the main capacitor (C1). Some coil output also goes through the brown wire to the spark advance module (SAM) mounted externally on the engine shroud. The conditioning circuit shapes the pulse, when then goes to what Kohler calls the charge pump. This circuit charges the main capacitor in a linear fashion related to engine speed.

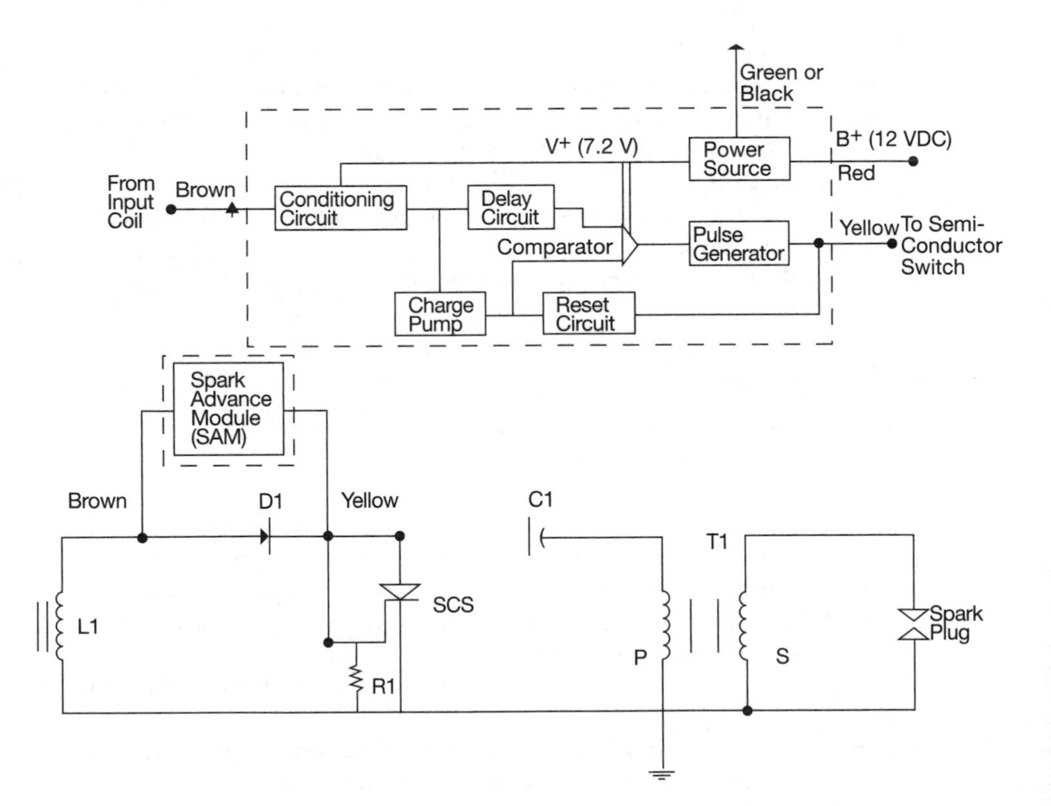

FIGURE 3-12. *Kohler Smart Spark system in block diagram.*

Both the charge pump and the delay circuit include capacitors. When the charge on the delay-circuit capacitor exceeds the charge on the charge-pump capacitor, the comparator fires the SCS (semiconductor switch). This action discharges the main capacitor through the primary side (P) of the transformer (T1), which is an ignition coil by another name. The resulting high voltage in the coil secondary (S) finds ground across the spark-plug electrodes.

Ignition timing depends upon how long it takes the two capacitors to reach parity. At low engine speeds, the delay-circuit capacitor is relatively slow to charge and ignition is delayed. As speed increases, the charge builds faster and the spark occurs earlier. The trigger pulse from the SAM discharges the capacitor in the reset circuit, clearing the decks for the next revolution of the crankshaft.

Troubleshooting Smart Spark

As for other CDI systems, replace the spark plug and, when damaged, the flywheel key. Make a careful visual examination of the external wiring, looking for loose connectors, bad grounds, and chaffed insulation. In some applications, these external circuits draw enough power from the CDI primary circuit to make starting difficult or impossible. Disconnect all equipment wiring before proceeding.

The spark advance module (SAM) needs 7.2 V to function. Verify that the battery has a full charge and make resistance checks of the ignition switch and associated wiring.

When a Smart Spark fails, the problem becomes one of deciding which module—ignition or SAM—to replace. Resistance tests of the ignition module, while by no means definitive, can help locate the problem. With the module at room temperature, disconnect the brown lead and test resistance from the wide connector tab to the coil laminations. Resistance should be 145–160 ohms. Remove the yellow lead and test resistance from the narrow tab to the laminations, which should be 900–1000 ohms. Finally, measure the resistance between the spark-plug terminal and the laminations. The meter should read between 3800–4400 ohms. Any out-of-range reading means that the ignition module should be replaced; otherwise, replace the SAM.

Magnetron

The Magnetron, used on Briggs & Stratton engines since the early 1980s, can be recognized by the single wire running from the coil to the kill switch. The most frequently encountered magnetos have two wires, one going to the switch and the other routed under the flywheel to the points (Fig. 3-13)

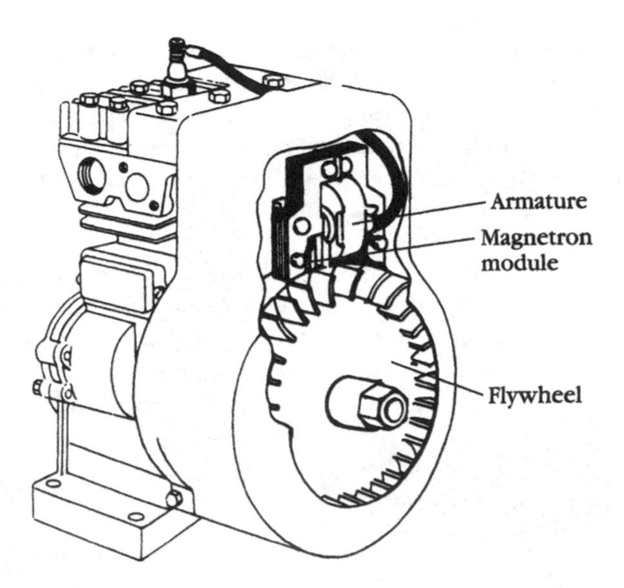

FIGURE 3-13. *The Magnetron mounts above the flywheel with a small-gauge primary wire going from the coil to the kill switch.*

Operation

The Magnetron incorporates a conventional ignition coil with primary and secondary windings, and a trigger coil piggybacked to it (Fig. 3-14). As flywheel magnets come into proximity with the coil, the moving field induces voltage in the trigger coil. This voltage causes the Darlington transistor—actually two paired transistors that function as a switch—to complete the primary circuit to ground. Further movement of the flywheel induces a current of about 3A in the primary, which saturates the secondary windings with magnetic flux.

As the flywheel turns further, magnetic polarity reverses. Trigger voltage changes polarity and the Darlington transistor switches OFF. Denied ground, current ceases to flow in the primary. The magnetic field surrounding the primary windings collapses in upon itself at near light speed. This collapse induces a high voltage in the secondary that goes to ground across the spark-plug gap.

The Magnetron advances ignition timing linearly with engine speed. At low speeds, the flywheel magnets must come to within close proximity of the trigger coil to induce the 1.2 V needed to activate the transistor. Higher speeds lower the magnetic threshold and ignition occurs earlier.

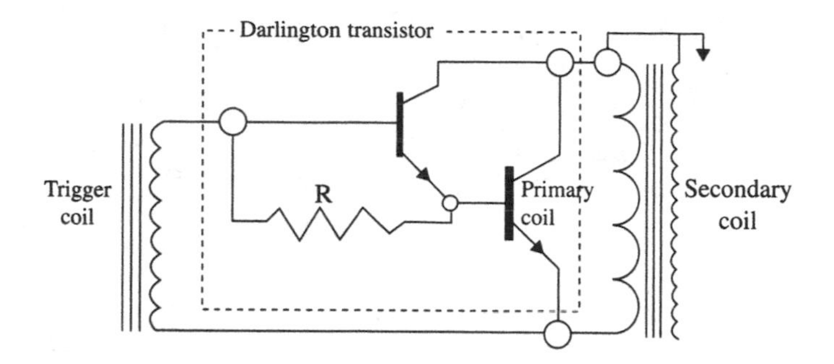

FIGURE 3-14. *Magnetron trigger circuitry appears to have been borrowed from automotive systems.*

Briggs stamps Magnetron coils with the date of manufacture. If the unit came with the engine, its date will be a month or so earlier than the engine build date.

Troubleshooting

Follow the same troubleshooting procedures as described for CDI systems with the caveat that the Magnetron is particular about spark plugs.

Service

Resistance readings between the spark-plug terminal and an engine ground should be between 3000–5000 ohms. Early models have a replaceable trigger module, which was a good feature since the transistors are vulnerable to failure from overheating.

Magneto-to-Magnetron conversion

The PN 394970 trigger-module converts magnetos found on many aluminum-block Briggs engines to Magnetrons. Doing away with the troublesome points and condenser is worth the $20 price of the kit, which includes installation instructions.

Magnetos

Although magnetos have been obsolete for decades, some manufacturers continue to use them. Consequently, anyone who works on small engines should come to terms with these sometimes-temperamental devices.

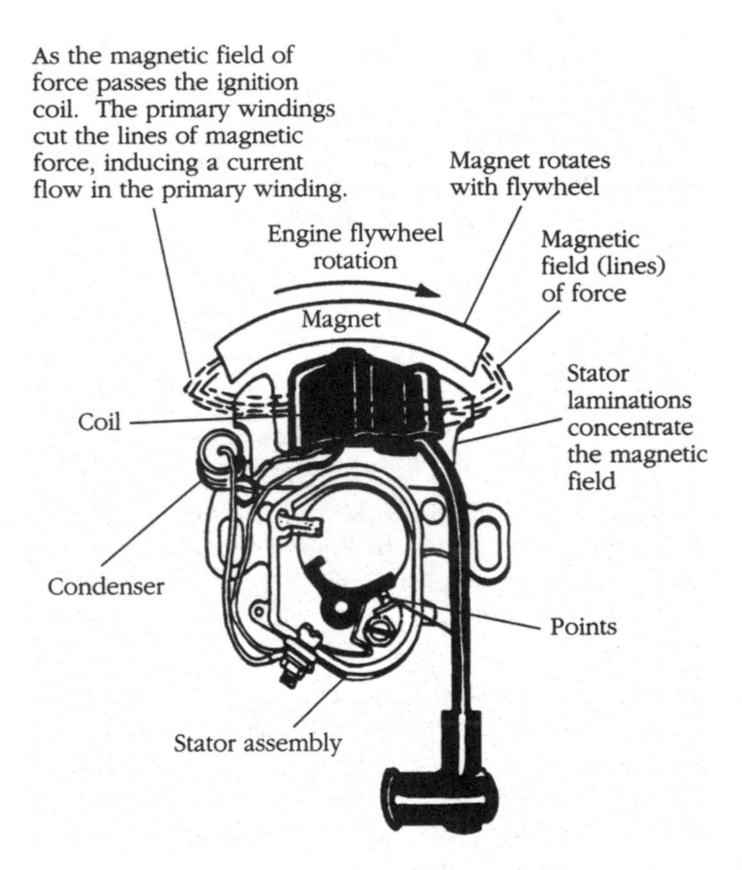

FIGURE 3-15. *An under-flywheel Phelan magneto. Elongated mounting-bolt slots on the stator assembly permit the unit to be rotated relative to the crankshaft for ignition timing. The drawing does not show the point cam.*

The unit shown in Figure 3-15 has all of its parts clustered under the flywheel which has magnets cast into its inner rim. Other designs mount the coil outside of the flywheel, in which case the rim magnets face outward. All of these designs fire every revolution, since the points actuate from a cam on the crankshaft. Vintage engines sometimes used camshaft-driven points, which eliminated the "phantom" spark on four-cycle engines.

Operation

The ignition coil, consisting of two electrically independent windings wrapped over a laminated iron core, is similar to those used on solid-state systems. The primary winding consists of about 200 turns of relatively heavy wire wrapped over the armature (Fig. 3-16). One end of the winding grounds to the coil armature; the free end goes to the moveable contact-point arm. The secondary winding consists of approximately 10,000 turns of hair-fine wire, wound on top of the primary, but insulated from it. One end of the secondary shares the same ground as the primary and the free end terminates at the spark-plug cable.

Figures 3-16 and 3-17 illustrate magneto operation about as well as drawings can. As the flywheel turns, a magnet sweeps past the coil to produce a voltage in the primary winding. When a moving magnetic field passes over a conductor, voltage appears in the conductor. When the contact points close, both ends of the primary circuit are grounded. Current then flows through this completed circuit.

The flow of primary current creates a strong magnetic field that saturates the secondary coil windings. Current flow, or the movement of electrons in a conductor, generates a magnetic field at right angles to the conductor.

Further movement of the flywheel cams the points apart; consequently, the primary circuit—now denied ground—no longer conducts. The magnetic field around the primary windings collapses in upon itself. This rapid collapse induces voltage in the secondary windings. Small-engine magnetos deliver 18,000–20,000 open-circuit volts. Nevertheless, like any ignition generator, a magneto produces no more voltage than necessary to

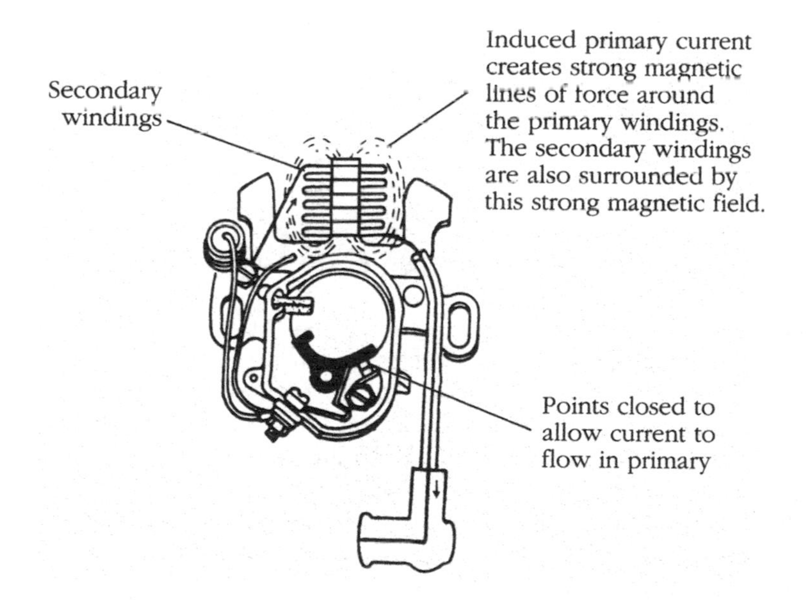

FIGURE 3-16. *With the points closed, current flows through primary windings to ground, saturating the windings with magnetic flux.*

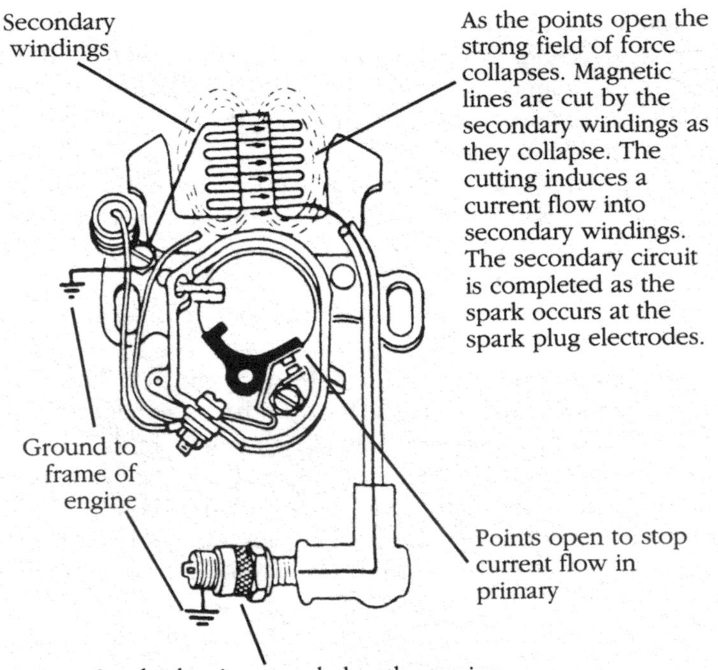

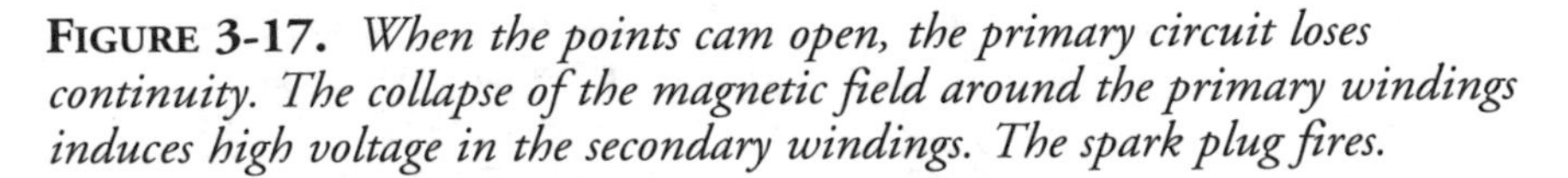

FIGURE 3-17. *When the points cam open, the primary circuit loses continuity. The collapse of the magnetic field around the primary windings induces high voltage in the secondary windings. The spark plug fires.*

overcome the resistance imposed by the spark gap. Operating voltages rarely exceed 6000 V.

The condenser provides temporary storage for electrons. The "hot" side of the condenser connects to the moveable point arm and through it to the primary winding. The other side of the condenser grounds to the engine through the metal case. When the points break, the condenser charges to store electrons that would otherwise find ground by arcing across the point gap. Milliseconds later, primary voltage diminishes enough to permit the condenser to discharge to ground through the primary winding. This backflow of electrons neutralizes primary voltage, speeding the collapse of the magnetic field and boosting secondary voltage.

Troubleshooting

Follow this procedure:

- Replace the spark plug with a known good one of the same type as originally specified.
- Replace the flywheel key if damaged.
- Replace the points and condenser.
- If there is still no spark, check the primary wiring that goes to the kill switch. Some applications use safety interlocks that open or short the primary circuit. See the "Interlocks" section at the end of this chapter.
- Finally, as a last and expensive resort, replace the ignition coil.

Contact points

Most magneto faults originate with the breaker points that sooner than later fail.

How points fail. New, out-of-the-box points can fail because of oxidation or the presence of oily fingerprints on the contact faces. Burnish with a business card.

Used point contacts should have a mottled appearance, but without the peaks and valleys associated with metal transfer. The tungsten contacts, bright as chrome on new point sets, turn slate gray in service.

Burnt points take on a darker color and the tip of the moveable arm sometimes shows blue temper marks. Patient filing can sometimes salvage a point set, but things go better if you opt for new parts.

Phelon and Wico point sets can short to ground through the moveable-arm spring, although this usually occurs as the result of an assembly error. The spring must not touch block metal. Point contacts can also become oil-fouled as a result of crankshaft seal failure or oil seepage through the plunger used on Briggs & Stratton and Kohler magnetos. Oil-fouling produces a splatter of carburized oil under the contacts.

All point sets loose gap as the rubbing block or plunger wears. A small amount of high-temperature grease on the cam helps to preserve the gap.

Servicing. Point assemblies for small engines come in two configurations. What we can call the "standard" configuration, found on most magnetos and on all battery-and-coil systems, consists of a moveable arm, a leaf-type point spring, and a fixed arm. The moveable arm bears against the point cam via a nylon or phenolic rubbing block. These point sets secure to the base plate, or stator, with one or two screws and alignment pins.

Remove all traces of oil from the mounting area and lightly lubricate the cam with high-temperature grease. Just a light smear around the full diameter of the cam is sufficient. Some cams lubricate from an oil-wetted wick, which can usually be reversed to present a fresh rubbing surface to the cam. Soak the wick in motor oil. Apply one or two drops of oil to the point pivot.

Install the point set, being careful not to contaminate the contact faces with fingerprints. Insert the locating pins on the underside of the assembly in holes provided in the stator plate. Tighten the electrical connection, being careful not to twist the moveable-arm spring into contact with ground. Lightly secure the hold-down screw(s).

Verify that contacts lie parallel and concentric with each other. Drawing A in Figure 3-18 illustrates full contact, with both point faces meeting squarely in the same plane. Drawing B shows the effects of misalignment. Snug down the mounting screws and correct misalignment by bending the *fixed* arm. Use long-nosed pliers or a proper bending bar, available from Tecumseh.

Adjust the point gap as follows:

1. A preliminary adjustment should be made to make and break the points as the flywheel turns. A screwdriver slot on the stationary-point bracket enables the gap to be widened or narrowed.
2. The flywheel is rotated until the points open to maximum. The rubbing block should be on the nose of the cam.
3. As far as the writer can determine, a 0.020-in. gap is standard for point sets of all models and vintages (Fig. 3-19). Bracket the gap specification by first inserting a 0.021-in. feeler-gauge blade between the contacts, followed with a 0.019-in. blade. The correct gap will produce a slight drag on the thicker blade and zero drag on the smaller.

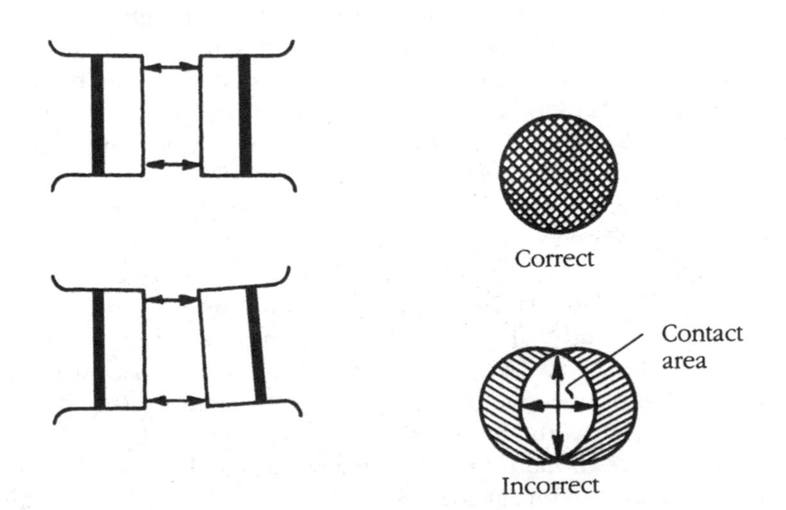

FIGURE 3-18. *Correct point alignment results in full contact and maximum service life. Adjust by bending the fixed point arm.*

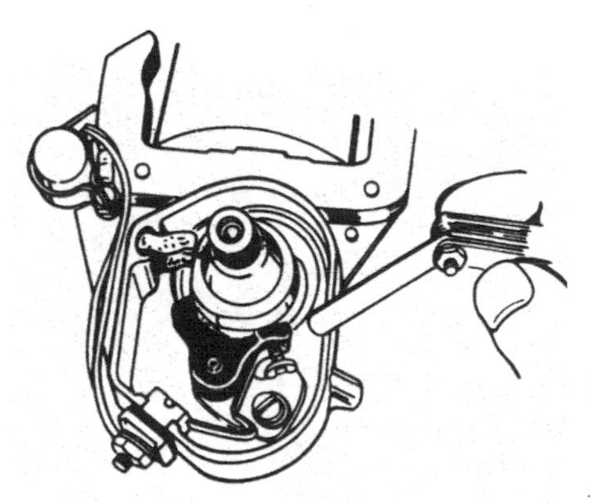

FIGURE 3-19. *Setting the gap for "standard" (fixed arm to ground) point sets.*

4. The hold-down screw(s) should be tightened, then check the gap, which will almost invariably have changed. Repeat the adjustment as many times as necessary, while attempting to anticipate the creep.
5. The contact faces should be burnished with a business card to remove fingerprints, oil, and oxidation.
6. A new condenser should be installed. Wipe off any oil from the mounting area and make certain the electrical lead clears the flywheel hub. Check that the moveable-arm spring has not twisted into contact with block metal.

B & S points. Briggs engine use point sets with the fixed contact "hot" and integral with the condenser. The moveable arm is grounded.
Install as follows:

1. Remove the flywheel and point-assembly cover (secured by two self-tapping screws), the condenser-clamp screw, and the breaker-arm screw (Fig. 3-20).
2. Note the lay of the parts:
   - The braided ground strap loops over the post that secures the breaker arm.
   - The open end of the point spring enters through the larger hole in the breaker arm and exits through the smaller hole. The closed spring end slips over the post on the stator plate.
3. Remove the coil and kill-switch wires from the condenser with the plastic spring compressor supplied with replacement point sets (Fig. 3-21). You can also use miniature water-pump pliers to "unscrew" the spring and release the wires.
4. Oil in the point cavity means a bad crankshaft seal or a worn plunger bore. The bore can be rebushed and reamed. Briggs supplied dealers with a reamer for this purpose.

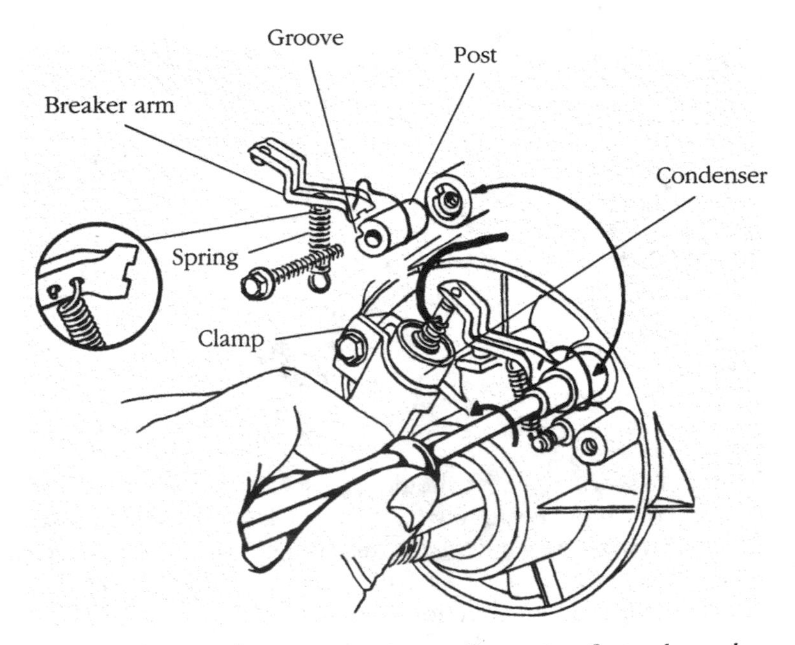

FIGURE 3-20. *Briggs & Stratton point configuration formerly used on light- and medium-frame engines. Note the spring orientation and the way the ground wire loops over the breaker arm and post.*

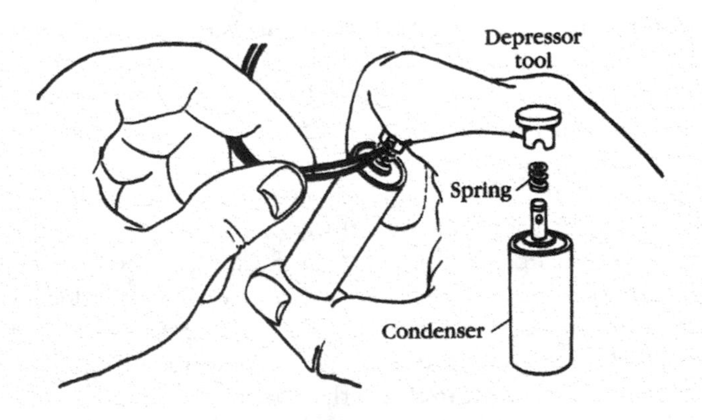

FIGURE 3-21. *Every small-engine mechanic should have a Briggs & Stratton ignition spring compressor in his tool box.*

5. The plunger is replaced if it measures less than 0.870 in. long. The grooved end of the plunger goes next to the point set.
6. Begin assembly by indexing the tubular breaker-arm post with its tab and routing the braided ground wire over the post as shown. Tighten the post screw.
7. Install the open end of the spring through the holes in the moveable arm as described above. Slip the closed end over the small post on the stator plate, seating the spring loop into the groove.
8. Grasp the movable arm and, pulling against spring tension, engage the end of the arm into the slot provided on the mounting post.
9. Using the tool provided with the replacement point set, compress the hold-down spring and insert the wires through the hole in the condenser terminal. Wires should extend about a quarter inch out of the terminal to make square contact with the spring.
10. Rotate the crankshaft to retract the plunger.
11. Install the condenser to bring the point faces together and lightly snug the clamp screw.
12. Rotate the crankshaft to fully extend the plunger and open the points.
13. Using a screwdriver, move the condenser as necessary to obtain a 0.020-in. point gap (Fig. 3-22).

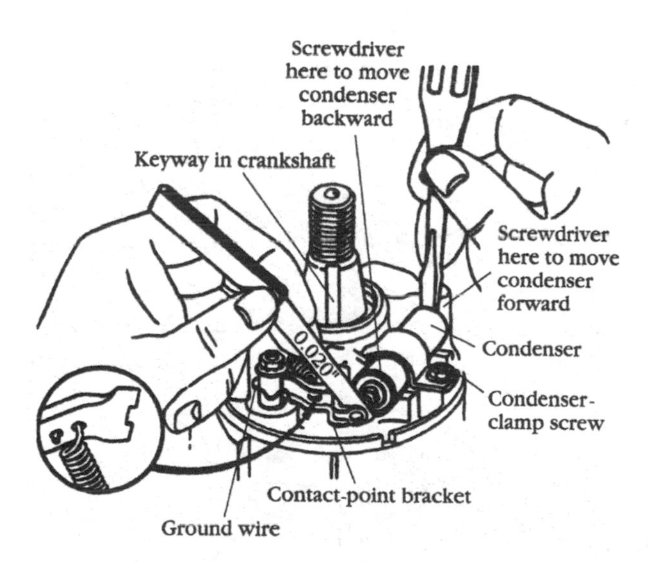

FIGURE 3-22. *To adjust Briggs & Stratton points, turn the crankshaft to bring the keyway in alignment with the point plunger, snug down the condenser-clamp screw, and set the gap at 0.020 in. Use a screwdriver to move the condenser as necessary. Tighten the clamp screw and recheck the gap.*

14. Tighten the condenser-clamp screw and check the gap, which will have moved out of specification. Redo the adjustment as necessary.
15. Burnish the contacts with a business card.
16. Install the flywheel key and flywheel. Lightly run down the nut or starter clutch and test for spark.

Armature air gap

The air gap, sometimes called the E-gap, is the distance the coil armature stands off from the rim of the flywheel. The narrower the gap, the stronger the magnetic field, and the more voltage induced in the coil. Ideally, the E-gap should approach zero. However, we need some clearance to accommodate main-bearing wear, thermal expansion, and production variations. Skid marks on the flywheel circumference mean the gap is too narrow.

The E-gap specification for all engines that the author is aware of falls within the range of 0.006–0.012 in. While a nonmagnetic feeler gauge could be used, most mechanics settle for a business card.

Follow this procedure:

1. Loosen the armature hold-down screws (Fig. 3-23).
2. Rotate the flywheel to bring the magnets adjacent to the coil armature. Lift the coil and insert the gauge. Flywheel magnets will pull the coil down snugly on the gauge.

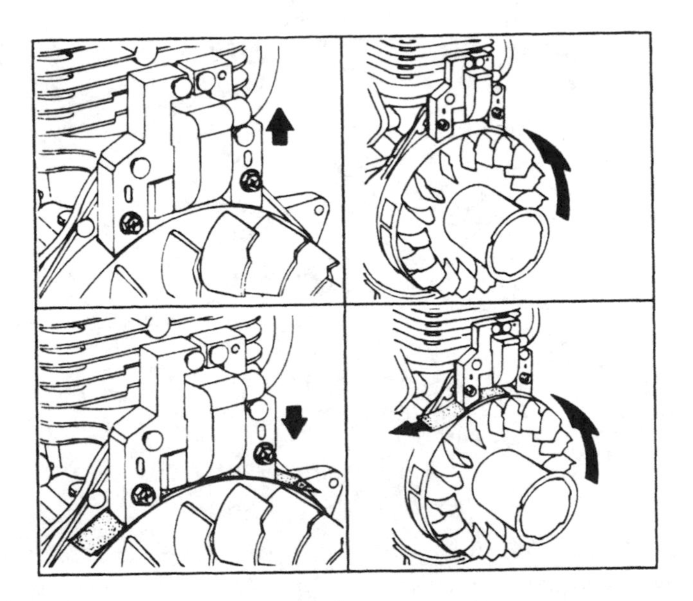

FIGURE 3-23. *Armature air-gap adjustment.*

3. Tighten the hold-down screws to 25 lb/in.
4. Rotate the flywheel to retrieve the tool.
5. The engine should be spun over several times to detect possible interference. At least 0.006-in. clearance is needed between the coil armature and all points on the flywheel rim.

Battery and coil systems

Battery and coil ignition, encountered on older Wisconsin, Kohler, and Onan engines, has several advantages, not the least of which is that the coil sees constant voltage. Owners of vintage magneto-fired engines might consider making the conversion. Moreover, by exercising a little ingenuity, one can update a b & c system with solid-state automotive components.

Operation

Figure 3-24 illustrates the system for a twin-cylinder engine. Both spark plugs fire simultaneously, with one cylinder on the compression stroke and the other on the exhaust stroke. When used on four-cycle engines, crankshaft-triggered ignition systems generate a "phantom" or superfluous spark every second revolution.

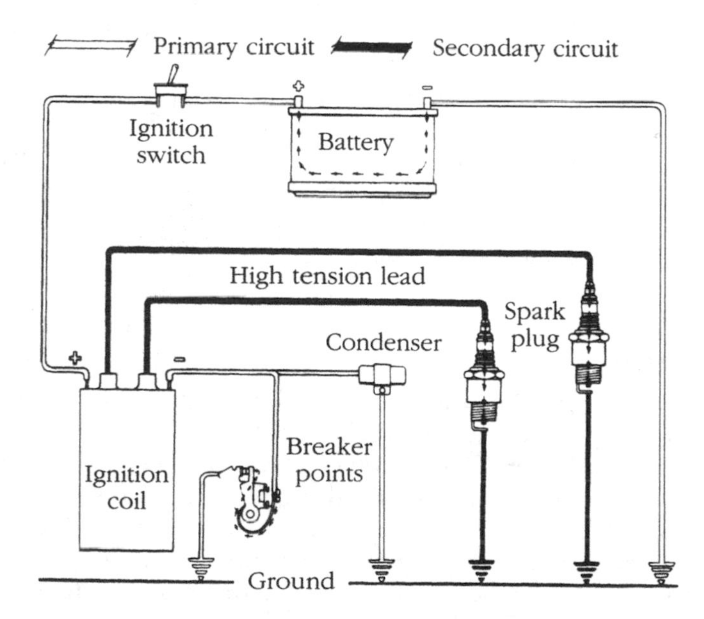

FIGURE 3-24. *Battery-and-coil ignition circuitry for twin-cylinder Kohlers.*

The primary side of the circuit consists of the battery, ignition switch, primary coil windings, breaker points, and condenser. Secondary coil windings connect to the spark plugs through dual high-tension leads. Both circuits ground to the engine.

The system functions like a magneto, except that battery voltage, rather than magnetic induction, provides power for the primary circuit.

Troubleshooting

First, verify battery condition, since most b & c ignition problems come about because of a weak battery. Points are the next most likely culprit, with the condenser and coil next. Chevrolet small-block V-8 point and condenser sets substitute for Kohler parts. The point gap for all applications is a tight 0.020 in.

With the switch ON, and the points open, check primary-circuit continuity with a test lamp or voltmeter. Connect the lamp across the moveable point arm and the fixed, or grounded, arm. The lamp should illuminate, although dimly because some voltage goes to ground through the primary coil windings. If no voltage can be detected, use the lamp to trace circuit continuity back to the battery. The ignition switch can fail outright or develop high resistance.

If the lamp illuminates when connected across the moveable and grounded points, turn the flywheel until the points close. The lamp will go out if the moveable contact grounds through the stationary contact. Should it remain lit, the points have oxidized.

When these tests turn up negative—that is, when primary voltage is present on the moveable arm with points open and absent with points closed—the problem can be assumed to be in the condenser or secondary circuit. Replace the condenser, if you have not already done so.

Check out the secondary circuit by substitution, replacing the least expensive parts first and reserving the coil for last.

Updating

Owners of battery and coil and magneto-fired engines, for which parts are scarce and expensive when found, might look into the ignition upgrade kit offered by Brian Miller. This kit, which combines automotive parts with a 12 V battery and a specially machined timing disc, adapts to virtually any small engine. Figure 3-25 shows the parts layout and Figure 3-26 illustrates features of the aluminum timing disc. The entry point for Brian's multiple Web sites is http://members.aol.com/ pullingtractor or he can be contacted by phone at 1-573-875-4033.

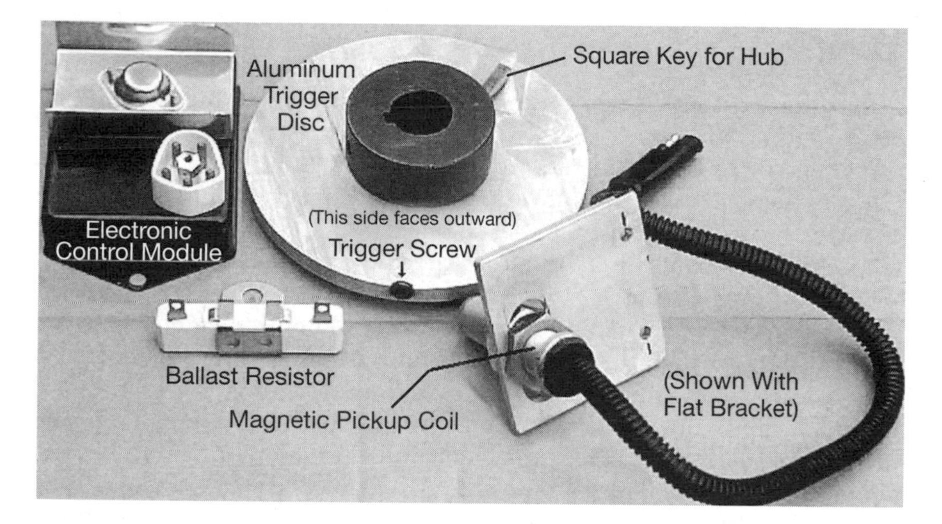

FIGURE 3-25. *Brian Miller's ignition upgrade kit delivers consistent sparks, with twice the duration of most factory systems, at speeds of 15,000 rpm and beyond. Because the triggering signal comes directly off the crankshaft, ignition timing has an accuracy of +/−0.1°. The tiny bit of slop reflects main-bearing clearance.*

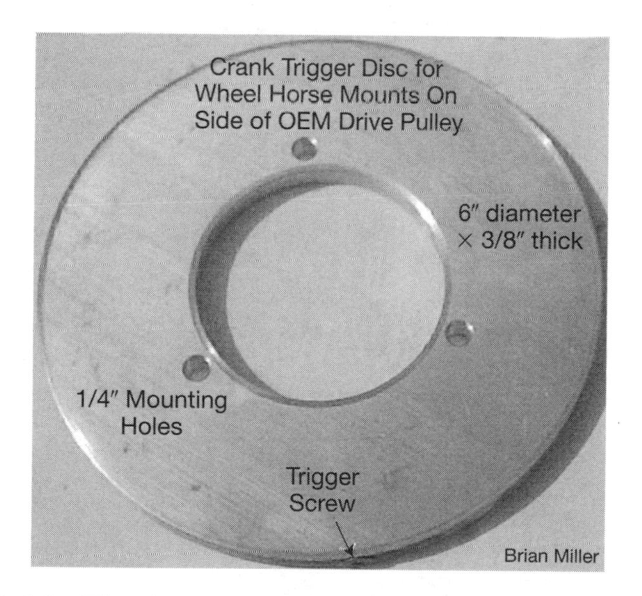

FIGURE 3-26. *The aluminum timing disc mounts on the pto end of the crankshaft and uses a precisely located steel screw to excite the trigger coil.*

Timing

Timing for most small engines cannot easily be changed in the field. Racers get around this handicap with offset flywheel keys that advance the timing a few degrees for improved mid-range torque.

Ignition systems with provision for timing adjustment fall into two groups. Those with contact points can be static timed with reference to point break; solid-state systems must be dynamically timed with a strobe light.

Static timing. Figure 3-27 illustrates stator timing marks for Wico and Phelon under-flywheel magnetos. Loosen the two hold-down bolts and rotate the stator to align the marks. While this falls short of real precision, it does at least restore timing to the factory setting.

Other engines time from a flywheel mark. When two marks are present, turning the crankshaft in the normal direction of rotation brings up the timing mark (sometimes labeled F or S) first. The second mark, about 20° of crankshaft rotation past the first, represents top dead center. When timed correctly, the contact points break (open) just as the flywheel timing mark indexes with its pointer.

An adjustable stator plate allows the point set to be moved a few degrees relative to the crankshaft. Nearly all small engines turn in a clockwise direction as viewed from the flywheel. Moving the point assembly counterclockwise advances the timing; moving the assembly clockwise delays ignition. Also, note that the point gap has a small, but significant effect on tim-

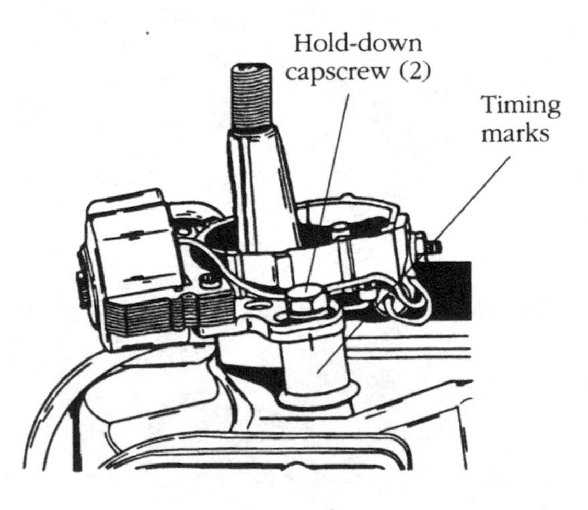

FIGURE 3-27. *Wico, Phelon, and most foreign magnetos time by rotating the stator to align with punch marks made during assembly. The marks are valid for the particular magneto/engine combination.*

ing. A larger than the specified 0.020-in. gap advances the timing by breaking the points earlier. A smaller gap retards timing.

Most mechanics use an ohmmeter connected between the kill wire and an engine ground to register point break. A piece of cigarette paper inserted between the contact faces can also be used to detect point break.

The ohmmeter reads zero or near-zero resistance with the points closed. A sudden jump in resistance signals that the points have opened. When timing is correct, the flywheel mark aligns with its pointer just as the points separate.

Honda G 150 and 200 engines hide their points under the flywheel, which complicates adjustment for those of us without access to the special factory tool. Install the flywheel over the key and run down the nut loosely. Determine point break with an ohmmeter. Note the position of the timing mark. Remove the flywheel and rotate the point box left to advance timing, right to retard. Replace the flywheel and retest. Several attempts will be necessary to synchronize point break with the mark.

Measuring piston movement btdc (before top dead center) gives more meaningful results than timing by marks. However, not all manufacturers provide the necessary specification. If you want to try this, you will need a dial indicator set up to mount in the spark-plug port (Fig. 3-28). For accuracy, remove all traces of carbon from the piston crown.

Locate top dead center, nudging the flywheel in ever-smaller increments until the piston pauses at the upper limit of its stroke. Zero the indicator and turn the flywheel in the normal direction of rotation until the points break. Note the indicator reading, which should be within one percent of the tim-

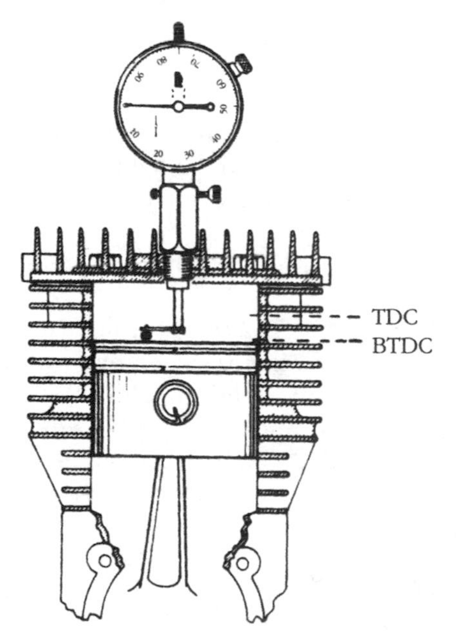

FIGURE 3-28. *Tecumseh supplies a dial indicator with a 14 mm adapter for engine timing.*

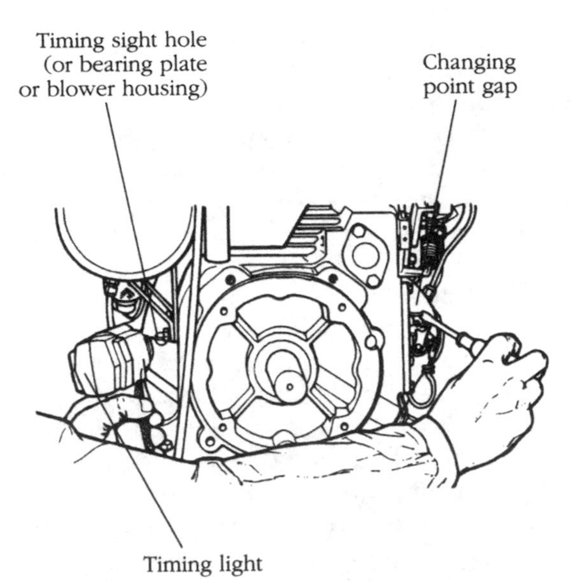

FIGURE 3-29. *The Kohler single-cylinder timing drill involves small adjustments to the point gap while the engine is idling and marks are frozen with a strobe light.*

ing specification. Correct by rotating the point assembly relative to the crankshaft. If that is not possible, vary the point gap by a few thousandths in either direction. Once you have set the timing, mark the flywheel and block so the engine can be timed dynamically in the future.

Dynamic timing. Any engine with external timing marks can be timed with a strobe light. Solid-state ignitions must be timed with a light.

Invest in a good timing light with an easily replaceable xenon bulb, an inductive pickup, and high-speed switching circuitry.

Warning: Do not look directly into the strobe. Xenon bulbs are bright enough to cause retina damage.

Engines with fixed advance time at idle speed (Fig. 3-29). The drill for automatic advance is more difficult to generalize. Some manufacturers (e.g., Sachs) provide two timing marks, one for idle rpm and the other for full advance. Others, such as Onan, provide a full-advance mark only, which means that timing must be accomplished at relatively high rotational speeds (Fig. 3-30). High-hour engines should have their advance mechanisms cleaned before setting the timing (Fig. 3-31).

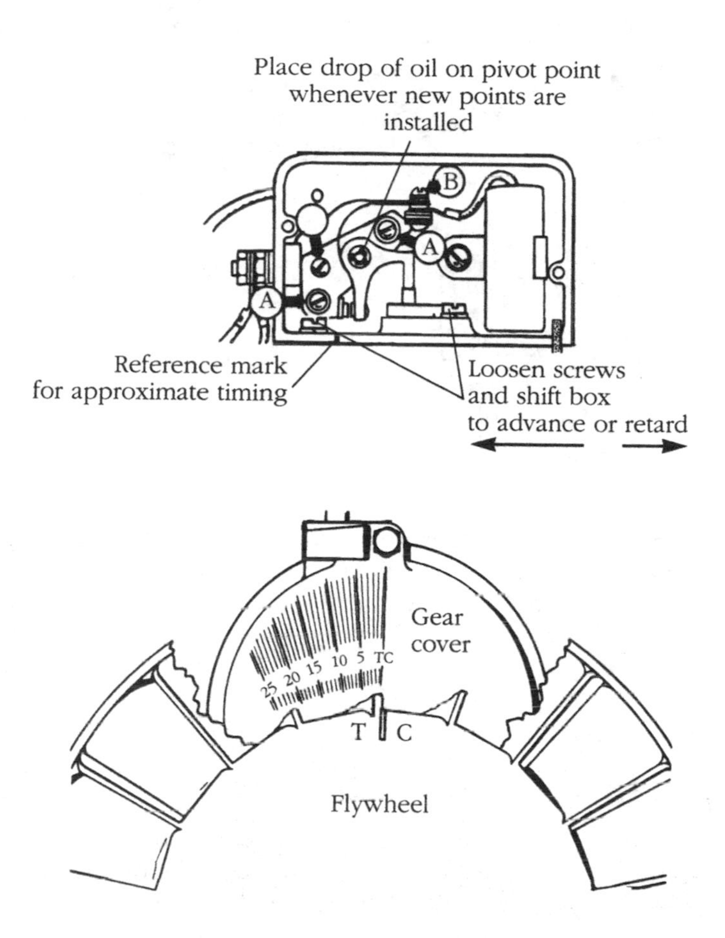

FIGURE 3-30. *Onan CCK and CCKA series engines time by moving the breaker box relative to the camshaft. TC stands for top dead center. However, Onan does not call out the timing mark. CCK engines fire 19° btdc, CCKA electric-start models with fixed advance are timed at 20° btdc and at 24° with auto advance.*

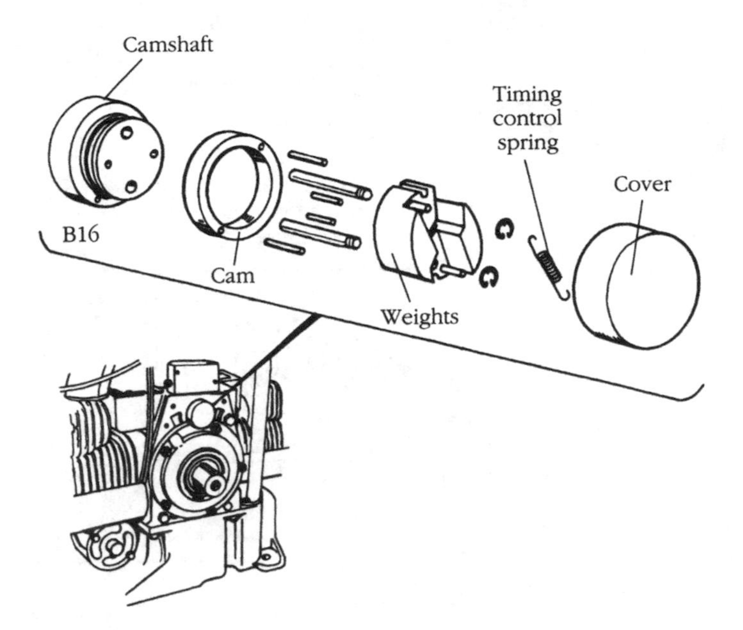

FIGURE 3-31. *Onan and other mechanical advance mechanisms should be inspected and cleaned prior to timing. Engines so-equipped must be timed by running at 1500 rpm or more.*

Interlocks

An interlock is an automatic switch that opens or grounds the primary circuit to protect the engine or operator. "Compliance" mower engines incorporate a grounding kill switch and flywheel brake that engage when the operator releases the handlebar lever (Fig. 3-32). Many modern four-strokes feature a float- or diaphragm-operated switch that shorts out the ignition primary when the crankcase oil level is low. Test these devices by temporarily removing them from the circuit.

Interlocks fitted to riding mowers, garden tractors, and other equipment vary in form and function. Some include a logic module. From a mechanic's point of view, interlocks fall into two categories: those that are normally open (NO) and those that are normally closed (NC). When energized, NO switches shunt primary current to ground (Figs. 3-32 and 3-33). NC switches open the primary circuit. In either case, the interlock denies ignition.

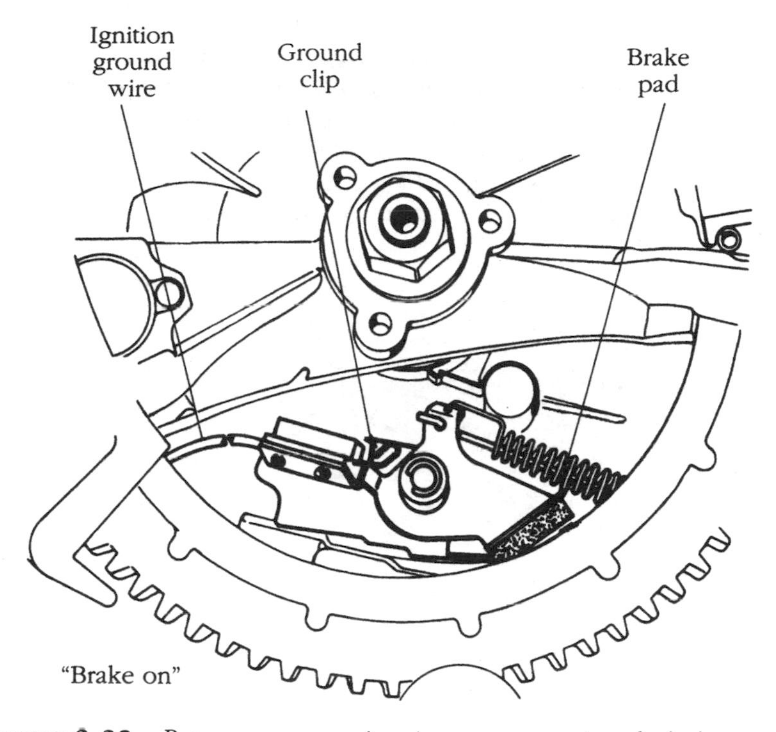

FIGURE 3-32. *Rotary mower engines incorporate an interlock that grounds the primary ignition circuit when the flywheel brake is engaged. Test the wiring with an ohmmeter or by disconnecting the primary lead at the coil. If ignition is restored, you can be sure that the switch or wiring has failed.*

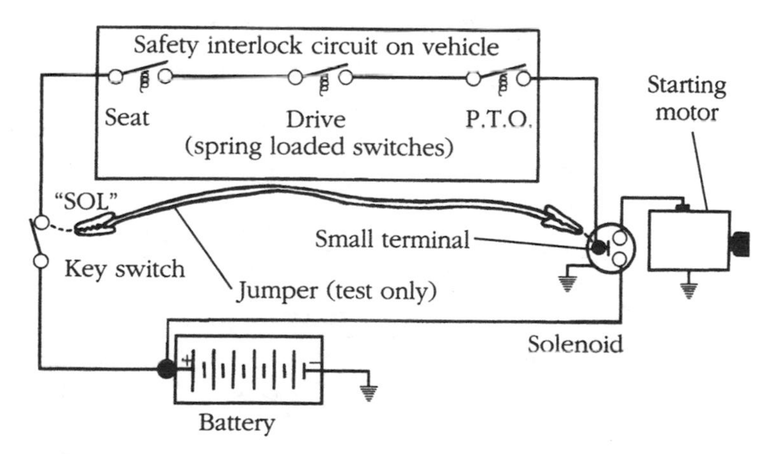

FIGURE 3-33. *A jumper wire is used to bypass suspect circuitry.*

Disconnect the associated wiring and test switch function with an ohmmeter. The meter should indicate near-zero resistance with contacts closed and infinite resistance with contacts open. A jumper can be used to shunt suspect components out of the circuit.

Caution: To avoid voltage spikes, do not open or make up wiring connections with the ignition key "on" or while the engine is running.

4

Fuel system

The fuel system consists of the carburetor, air cleaner assembly, fuel tank, and optional components such as a filter, shutoff valve, and fuel pump.

Tools and supplies

In addition to standard mechanic's tools, you will need the following:

- Flat-bladed screwdriver ground to fit carburetor jets. Damaged screw slots affect flow through the orifice.
- Magnifying glass.
- Carburetor cleaner. Aerosol cleaners suffice for local deposits; seriously contaminated carburetors should be stripped of all nonmetallic parts and immersed in a chemical cleaner such as Wynn's.
- Pipe cleaners are handy for cleaning fuel passages, but should not to be inserted into jets. If a clogged jet cannot be blown out, clear the orifice with something soft, like a broom straw.
- A source of compressed air.
- Serious DIY mechanics would do well to purchase a Walbro PN 500–500 tool kit, which includes diaphragm-lever gauges and a chisel and punch assortment for removing and installing Welch plugs. Walbro also supplies a tool for cycling diaphragm-type carburetors.

Engine condition

Before getting started, we need to say something about the ability of the engine to draw a vacuum. Manifold vacuum must be present for a carburetor to function.

In general, we can assume that any four-cycle engine that develops cylinder compression also draws a vacuum in the inlet tract. A vacuum will be present unless there is a massive leak across the carburetor mounting-flange gasket or a malfunction that prevents one or both valves from opening.

The situation becomes a little more complicated for two-strokes. Fuel delivery depends upon the ability of the crankcase to hold pressure. Fortunately, loss of crankcase integrity is a rare malady, nearly always associated with leaking crankshaft seals in high-hour engines. Some novice mechanics have never encountered the problem, or if they have, did not recognize it.

"Yes, Mam... there's something wrong with the carburetor, but the engine runs okay if you leave the choke on."

The definitive test for crankcase integrity is to make up an adaptor plate to the carburetor mounting flange and pressurize the case to 5 or 6 psi. Observe the pressure drop on a gauge. Some air will leak past the piston, but failure to hold any pressure means that the seals have failed. Other, far less likely possibilities are a leaking reed valve or a defective crankcase casting.

A substitute for this test is to squirt a little carburetor cleaner into the spark-plug port or carburetor intake. If the engine starts and runs for a few seconds, we can be reasonably confident that the seals are good and that the problem lies elsewhere.

Troubleshooting

Do not rush into the job. Test ignition system output with a spark-gap tool and install a known-good spark plug of the specified type. Drain the tank and refill with fresh gasoline or premix. Clean foam air filters and replace paper air and fuel filters. Verify that carburetor mounting bolts are tight and that the choke, whether manual or automatic, closes fully on a cold engine.

Depending upon how the internal fuel level is regulated, small-engine carburetors fall into three groups—float, diaphragm, and suction lift. Float carburetors can be recognized by the conical float chamber under the main casting (Fig. 4-1); diaphragm carburetors by the absence of a float chamber (Fig. 4-2); and suction-lift carburetors by the way they are piggybacked on top of the fuel tank (Fig. 4-3).

No fuel delivery

Zero fuel delivery is obvious because the spark-plug tip remains resolutely dry after prolonged cranking. Heroic efforts might oil the tip on four-cycle engines, but the characteristic odor of gasoline will be absent. The carburetor bore, visible when the air cleaner is removed, will be dry or, at best, damp. Spray carburetor cleaner into the spark-plug port. If the engine briefly comes to life, the problem is fuel starvation. Backing out the jet adjustment screws (when present) may be enough to restore fuel flow.

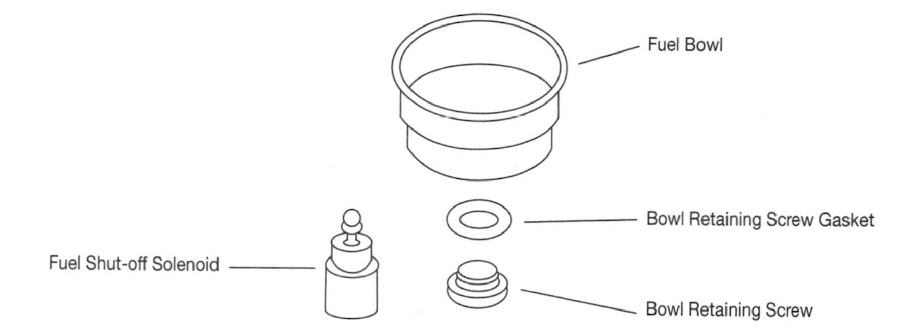

FIGURE 4-1. *A float-type carburetor with optional solenoid fuel cutoff used on several Kohler engines.*

Float-type carburetors rarely fail to pass fuel unless contaminated by exposure to stale gasoline or water. The problem is more often upstream of the carburetor. Replace the fuel filter if you have not already done so, check the tank screen and the optional fuel pump. Cracking the fuel line at the carb connection should yield a dribble of gasoline from gravity-fed systems. Pump-fed systems must be activated by cranking the engine.

The solenoid-operated fuel shutoff valve found on some Walbro and Nikki carburetors requires a minimum of 7.3 V to function (Fig. 4-1). These valves can fail outright, in which case the engine will not start, hang partially open to lean the mixture, and cost power. Test by replacing the valve with the standard brass float-bowl fastener.

Lack of fuel delivery is a frequent complaint with diaphragm-type carburetors. Replace the diaphragm and remove any trace of varnish from the needle and its supply passages with aerosol carburetor cleaner.

Many diaphragm carburetors have a second diaphragm that functions as a fuel pump. The plastic line connecting the pump diaphragm chamber with the crankcase may lose resiliency and leak air at the connections. The pump diaphragm, while not as troublesome as the metering diaphragm, should also be replaced.

The check valve on Vacu-Jet siphon-feed carburetors tends to stick shut during extended layups. Insert a fine wire, such as from a wire brush, through the screen at the bottom of the tube and gently dislodge the ball.

Flooding

Any carburetor can flood and dribble raw gasoline from the air cleaner or overflow tube if over-primed or over-choked. However, self-induced flooding is a serious and potentially hazardous malfunction.

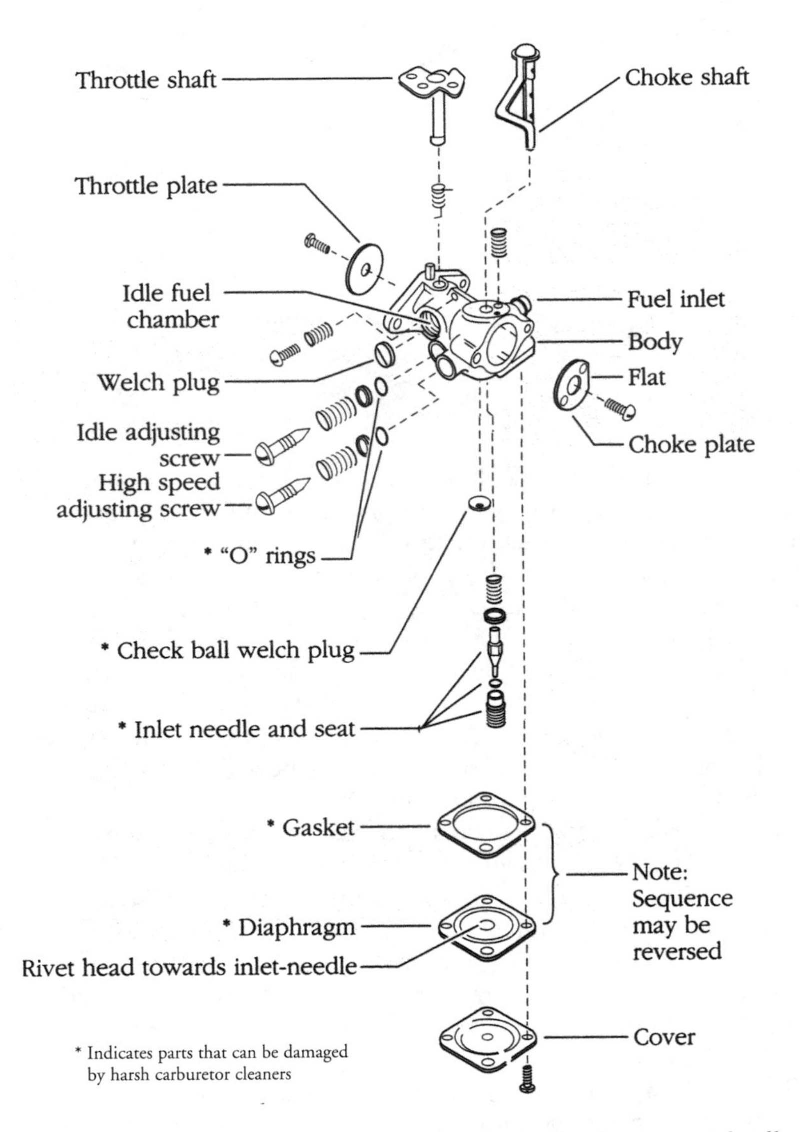

FIGURE 4-2. *A Tecumseh pre-emission diaphragm carburetor with idle and high-speed mixture adjustment screws.*

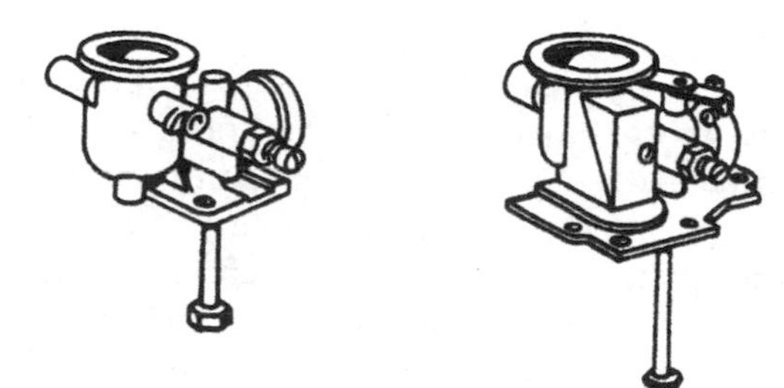

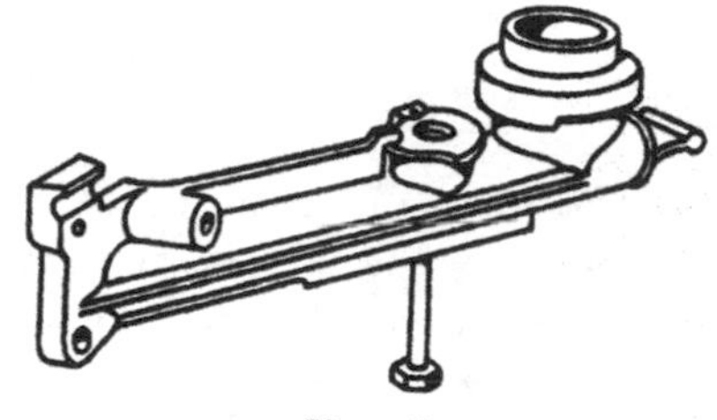

Vacu-Jet

A

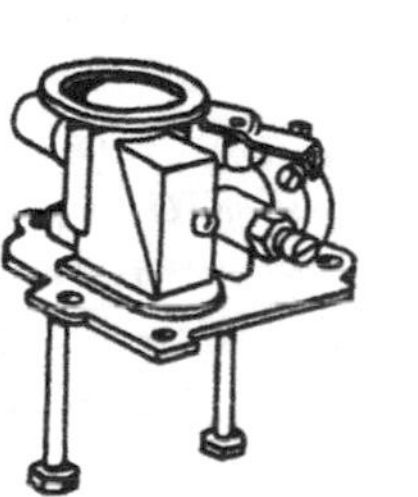

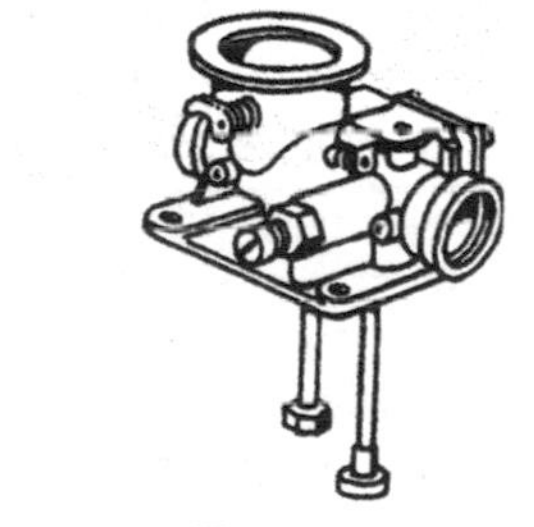

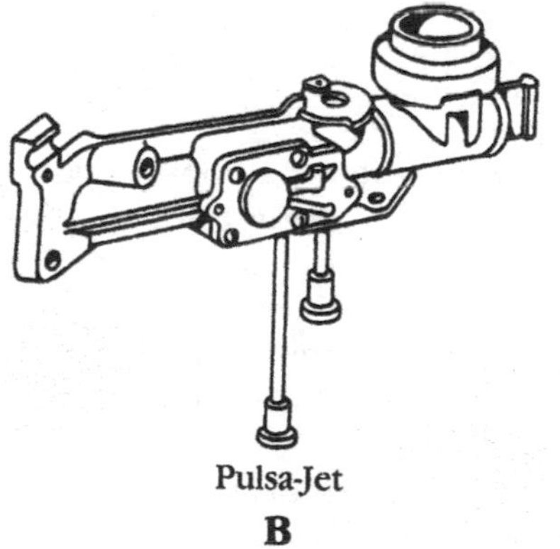

Pulsa-Jet

B

FIGURE 4-3. *Briggs & Stratton Vacu-Jet (A) and Pulsa-Jet (B) carburetors.*

I have never seen a suction-lift carburetor flood spontaneously; diaphragm carburetors can flood, but the condition is extremely rare. Float-type carbs flood regularly—a condition that can usually be traced to a defective inlet needle and seat. Another source of flooding is loss of float buoyancy. Hollow plastic floats give no problem, but solid plastic floats can become weighed down with absorbed fuel. Metallic floats eventually leak, a condition revealed by vigorously shaking the float and listening for fuel slosh. Some carburetors are plagued by hung floats that drop out of contention when fuel in the chamber evaporates.

Refusal to idle

All carburetors: Refusal to idle can be caused by restricted throttle-plate movement, an obstruction in the idle circuit, or a vacuum leak downstream of the carburetor. Air leaking past a failed gasket or o-ring seal at the carburetor mounting flange increases idle rpm.

Failure of the throttle plate to close can be caused by the following:

- Idle rpm set too high. Adjust the throttle stop screw as necessary. But note that small engines do not "Cadillac." If a four-cycle engine can be persuaded to tick over at a few hundred rpm, it may throw a connecting rod when accelerated abruptly. Two-strokes pop and sputter at idle, a condition caused by poor scavenging and/or fuel puddling in the crankcase. A mechanic can do little about these design flaws.
- Maladjusted throttle cable. Loosen the clamp screw and adjust the Bowden cable to restore idle.
- Binding throttle linkage or throttle-butterfly shaft. The latter condition can sometimes be encountered on freshly painted engines.
- Malfunctioned or maladjusted governor. See the "Governor" section at the end of this chapter.

The most common malfunction is an obstruction in the idle-speed circuit, usually at the jet or at the discharge ports drilled in the side of the carburetor bore adjacent to the throttle blade. Backing out the idle-mixture adjustment screw compensates for a partial blockage; removing the screw and blowing out the circuit with compressed air either clears the jam or compacts it further. Ultimately, you will need to dismantle the instrument for a thorough cleaning.

Diaphragm carburetors idle erratically if the adjustment lever (illustrated in the following section) is set too high or the wrong metering spring is installed.

Refusal to run at high speed

All carburetors: Failure to attain rated rpm has several causes, including loss of tension in the throttle-return spring, a maladjusted throttle cable, binding throttle linkage, or a governor malfunction. Stretched throttle-return springs rate high on the list of possibilities.

Warning: Do not shorten or otherwise modify malfunctioning throttle springs. Replace the spring with the correct part number for the application. Otherwise, the engine may overspeed.

Vacuum leaks downstream of the throttle plate also cost power and loss of rpm. Check that carburetor-mounting bolts are tight. If the carburetor is loose, the flange gasket or o-ring is probably damaged and should be replaced. As indicated previously, two-cycle engines can leak air past the crankcase seals, a condition that denies full throttle unless the choke valve is partially closed.

Insufficient fuel delivery can result from internally collapsed fuel lines or partial stoppages of fuel filters and screens. As far as the carburetor is concerned, the problem is associated with the high-speed circuit. Float-type carburetors with their main jets positioned low in the bowl often clog. Diaphragm carburetors go lean because of loss of resiliency of the metering diaphragm or a varnish accumulation on the inlet needle.

Adjustable main jets cover a multitude of sins, including partially blocked high-speed circuits, restrictive air cleaners, and vacuum leaks. But large adjustments should not be necessary. A healthy carburetor holds adjustment for the life of the machine with only an occasional tweak.

Black smoke, acrid exhaust

These symptoms point to an excessively rich mixture. The problem can be caused by improper mixture adjustment, a clogged air cleaner, or a choke valve that does not fully open. Weeping inlet needles and seats on float-type carburetors have a similar effect.

Stumble during acceleration

Hesitation when the throttle is snapped open can usually be cured—or masked—by enriching the idle or high-speed mixture or both. For better idle quality, newer diaphragm carbs sometimes employ an accelerator pump. As the throttle pivots open, the pump delivers a shot of fuel into the carburetor bore. Construction varies—some designs exploit crankcase pressure fluctuations to excite the main metering diaphragm or a second, smaller diaphragm. Others use a spring-loaded brass pump plunger tripped by the throttle. These systems work better than one might expect, so long as the tiny and highly convoluted passageways are clean.

Hot start difficulties

Ignition-coil failures cause most hot-start difficulties, with vapor lock coming in a distant second. Once you have verified that spark is present, make sure that carburetor insulators and muffler heat shields are in place. The use of highly volatile winter-grade gasoline in hot weather or a rich carburetor setting exacerbates the problem.

Removal and installation

The carburetor bolts to the inlet flange or to the top of the fuel tank. When dealing with gravity-feed systems, the fuel supply must be shut off with Vise-Grips clamped on the hose if a shutoff valve is not present. It is good practice to clean the external surfaces of the carburetor before removal.

Remove the air cleaner and set the gasket aside. The governor mechanism must be disengaged from the throttle arm without doing violence to the springs and wire links. Detach the springs, but leave the wire links connected for now. Most springs have open-looped ends and can be coaxed out of their mounting holes with a gentle twist. When multiple holes are provided, as on a governor arm, note the hole used. Linkages can be quite complex and you may want to make a drawing.

Next, remove the carburetor mounting screws. Holding the carburetor in one hand, twist and rotate it out of engagement with the governor link(s).

Installation is the reverse of assembly; that is, hook up the wire throttle links before the carburetor is bolted down and can still be manipulated.

Repairs

So long as the aluminum or pot metal casting has not oxidized, a few relatively inexpensive parts can put nearly any carburetor back into service.

Lay out the parts in order of disassembly on a bench covered with clean paper. Do not disturb:

- Welch (expansion) plugs unless necessary to clear idle-speed or other critical circuits. Removal and installation of these plugs is described later in this chapter under "Diaphragm carburetor service."
- Throttle and choke plates. Dismantling these components invites difficulties with stripped screws and plate alignment upon assembly. The

only time these parts should be disturbed is to replace worn bushings and throttle shafts on carburetors that support these repairs.
- Pickup tubes on Briggs & Stratton suction-lift carburetors.
- Pressed-in parts. Many parts that were formerly threaded, such as inlet fittings, main nozzles, and jets, are now pressed in and should be left in place.

Slightly dirty carburetors clean up with lacquer thinner or aerosol carburetor cleaner and compressed air. Blow out the passages in the reverse direction of flow.

Caution: Do not insert a wire or other hard objects into jet orifices.

The only way to deal with a really dirty carburetor is to remove all soft (plastic and elastomer) parts and soak the metallic parts in a chemical cleaner for 20 minutes or so. But these powerful cleaners cannot be used on nylon-bodied suction-lift carburetors or on diaphragm carburetors that have non-removable plastic parts. Nor can cleaning restore damaged metal surfaces. A carburetor that looks like the one shown in Figure 4-4 belongs in the trash barrel.

FIGURE 4-4. *This is what stale, water-contaminated gasoline does to a carburetor.* Robert Shelby

A rebuilt kit, available from the engine dealer, should include new gaskets, inlet needle and seat, diaphragms, Welch plugs, and o-rings. Some kits contain a float-height or diaphragm-lever gauge. If not, query the dealer for these specifications.

Walbro float-bowl gaskets are one-shot affairs that expand when wetted with gasoline. The original gasket sometimes shrinks enough when dry to be reused, but anyone who works on small engines should keep several extras on hand.

Carburetor types

As indicated earlier, small-engine carburetors vary by the way fuel is admitted. Float-type carburetors work on the same principle as toilet tanks; diaphragm-type carburetors regulate fuel entry with a flexible membrane; and suction-lift carburetors work like a flit gun, drawing fuel through a pickup tube that is exposed to a vacuum on its upper end.

Float-type carburetor operation

The regulating mechanism consists of an inlet valve, also known as the needle and seat, and a plastic or hollow brass float (Fig. 4-5). As fuel in the bowl is consumed, the float drops, allowing the needle to fall away from its seat. Fuel enters the bowl until the float rises and closes the valve, an action that occurs several hundred times a minute at full throttle. In order to allow the engine to operate off the horizontal, the fuel pickup is at the center of the bowl.

Suction-lift and diaphragm carburetors evolved from these float-type instruments and share common features with them. The paragraphs that follow describe these features.

High-speed circuit

The main point of fuel entry is at the venturi, a necked-down section of the bore. When the air stream encounters this restriction, it accelerates and simultaneously loses pressure (Fig. 4-6).

Fuel, under atmospheric pressure, moves from the float bowl through the main jet and nozzle (or, as the drawing has it, the main pickup tube) to discharge into the low-pressure area created by the venturi (Fig. 4-7A). The main jet and its associated components make up the high-speed circuit—"high-speed" because this circuit flows only when the throttle is open. Most high-speed circuits shut down at about one-quarter throttle.

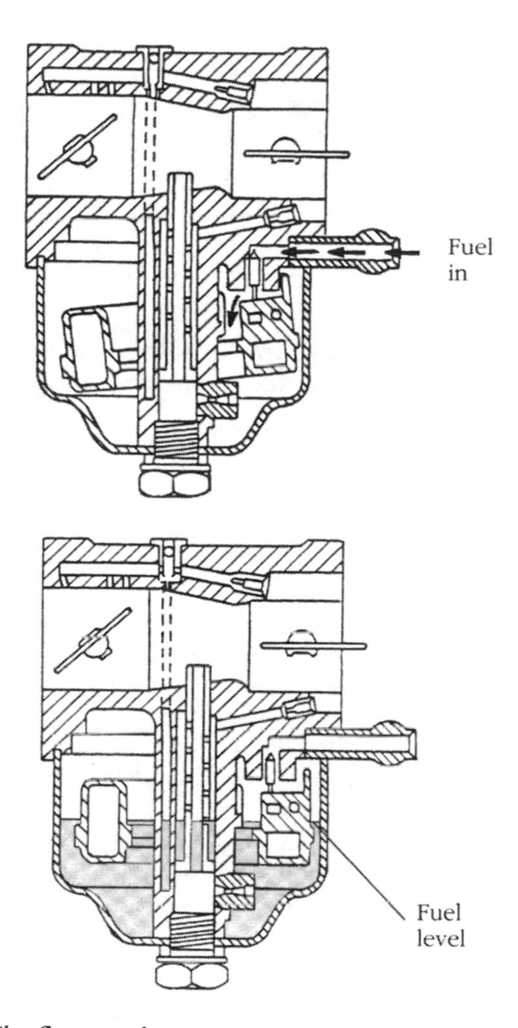

FIGURE 4-5. *The float mechanism maintains a preset fuel level in the float chamber and above the main jet. Failures most often involve the inlet needle and seat, a mechanism that can be defeated by a speck of dirt.*

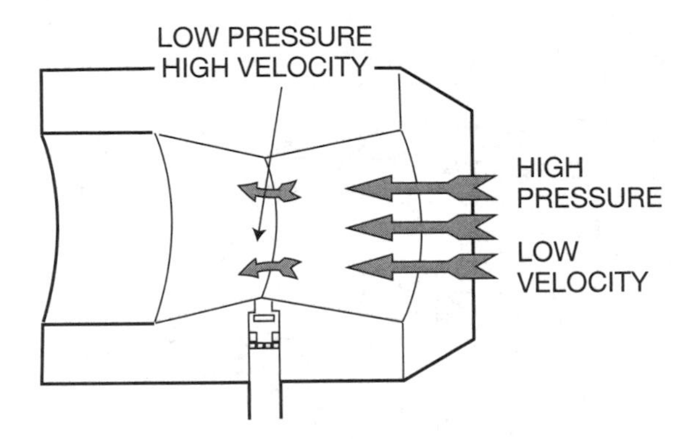

FIGURE 4-6. *Fuel discharges through the nozzle (shown on the lower part of the drawing) into the low-pressure, high-turbulence zone created by the venturi.* Walbro Corp.

Idle circuit

A closed or nearly closed throttle plate represents a major restriction or, if you will, a kind of crude, unstreamlined venturi (Fig. 4-7B). Very low pressures develop near the trailing edge of the plate. Idle-circuit ports discharge fuel into this depression. The port nearest the engine, the primary idle port, flows when the throttle rests against its stop. Secondary idle ports (Fig. 4-7C) come onstream as the throttle cracks open to smooth the transition to the high-speed circuit.

One of the effects of emissions regulations has been to eliminate adjustable main jets on all carburetors and adjustable idle-speed jets on many. The example illustrated features adjustable jets.

Cold-starting aids

Virtually all carburetors employ some mechanism for richening the mixture during cold starts. Traditionally this has been done with a choke valve mounted upstream of the venturi. When the choke is closed, all circuits come under vacuum and flow.

String trimmers and other portable tools often employ a primer—a miniature pump—in lieu of a choke.

Float-type carburetor service

Figure 4-8 lists things to look for whenever a float-type carburetor comes in for service.

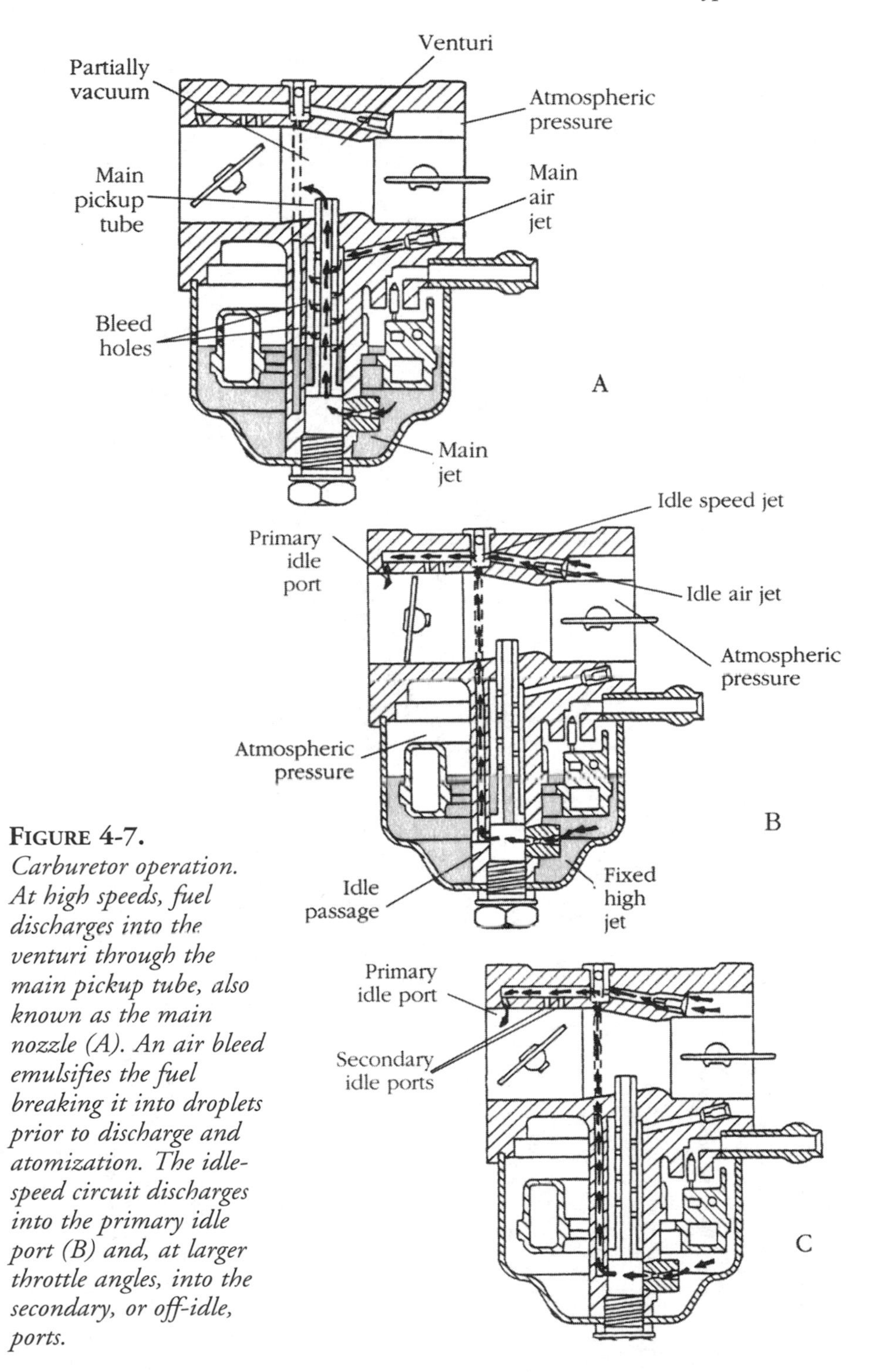

FIGURE 4-7.
Carburetor operation. At high speeds, fuel discharges into the venturi through the main pickup tube, also known as the main nozzle (A). An air bleed emulsifies the fuel breaking it into droplets prior to discharge and atomization. The idle-speed circuit discharges into the primary idle port (B) and, at larger throttle angles, into the secondary, or off-idle, ports.

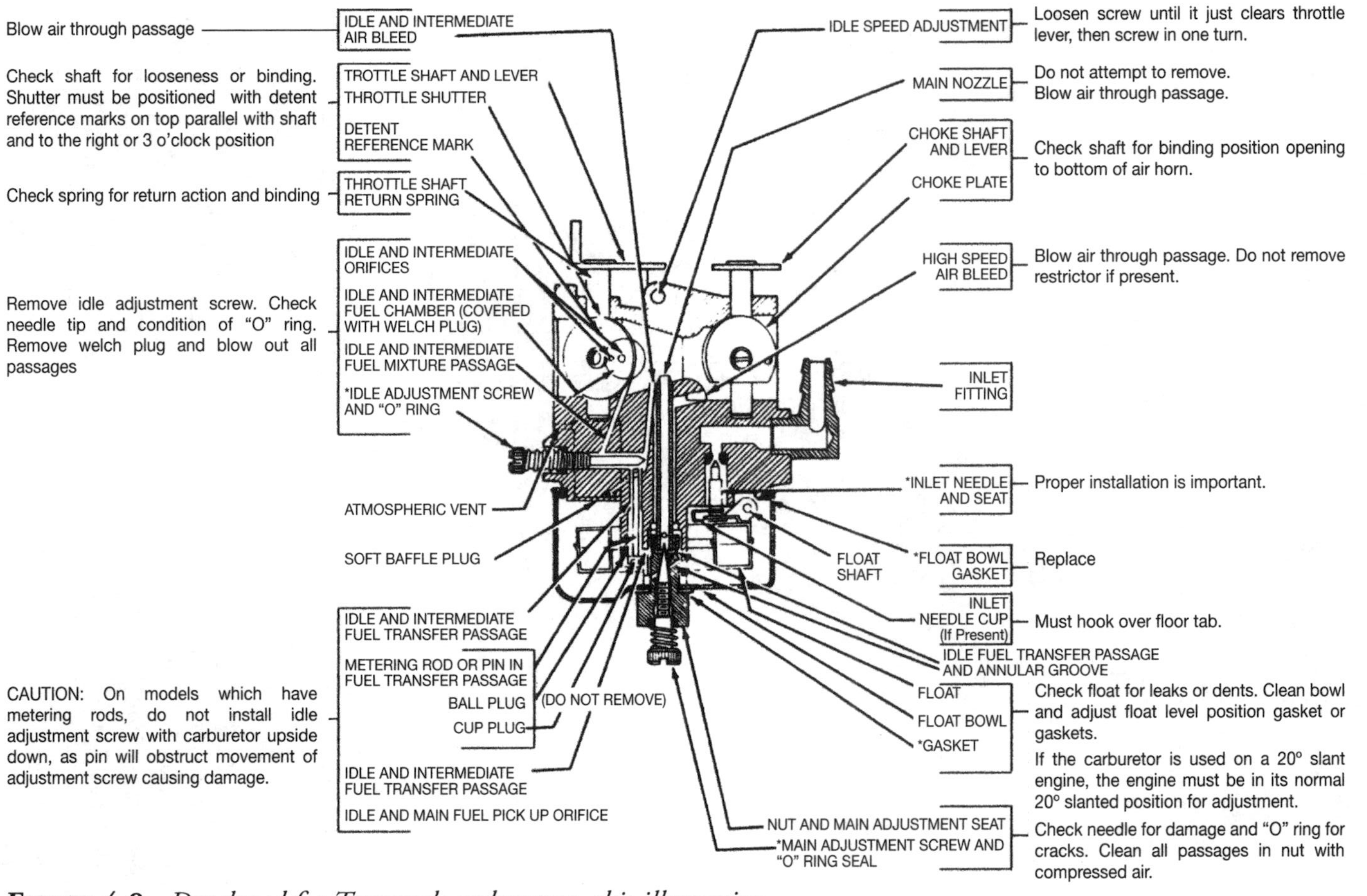

FIGURE 4-8. *Developed for Tecumseh carburetors, this illustration has general application for other float-type units.*

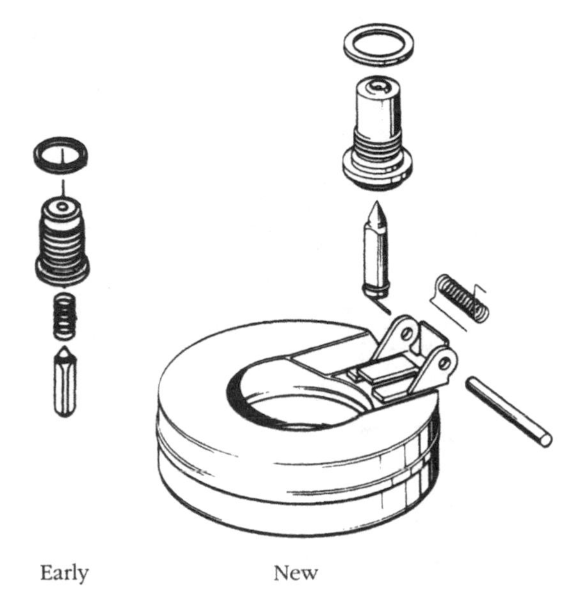

FIGURE 4-9. *The all-steel inlet valve on the right evolved into the elastomer-tipped valve on the left. Both examples include dampening springs to cushion float action.*

Needle and seat. Chrome-steel needles and brass seats, shown on the right in Figure 4-9, were the norm back in the days of the Carter Model N. It was considered good practice to give these needles and light rap with a wrench upon assembly. The newer Viton-tipped needles, like the one on the left, do not tolerate such rough treatment. Care must be exercised to avoid deforming the needle when adjusting the float height.

Several U.S. manufacturers substitute an elastomer disc for the brass seat. If you have one of these carburetors, install a fuel cutoff valve at the tank or in the fuel line; otherwise, you are liable to wake up one morning with a garage full of gasoline. These things leak. Replace the seat as shown in Figure 4-10. Buy several, because arriving at the proper installation force can be tricky. Too little force and the seal leaks around its OD, too much and the orifice distorts.

Various combinations of float dampener springs, needle buffer springs, and needle spring clips are shown back in Figure 4-9 and, in further detail in Figure 4-11. Incorrect installation can cause the float to hang and the carburetor to flood.

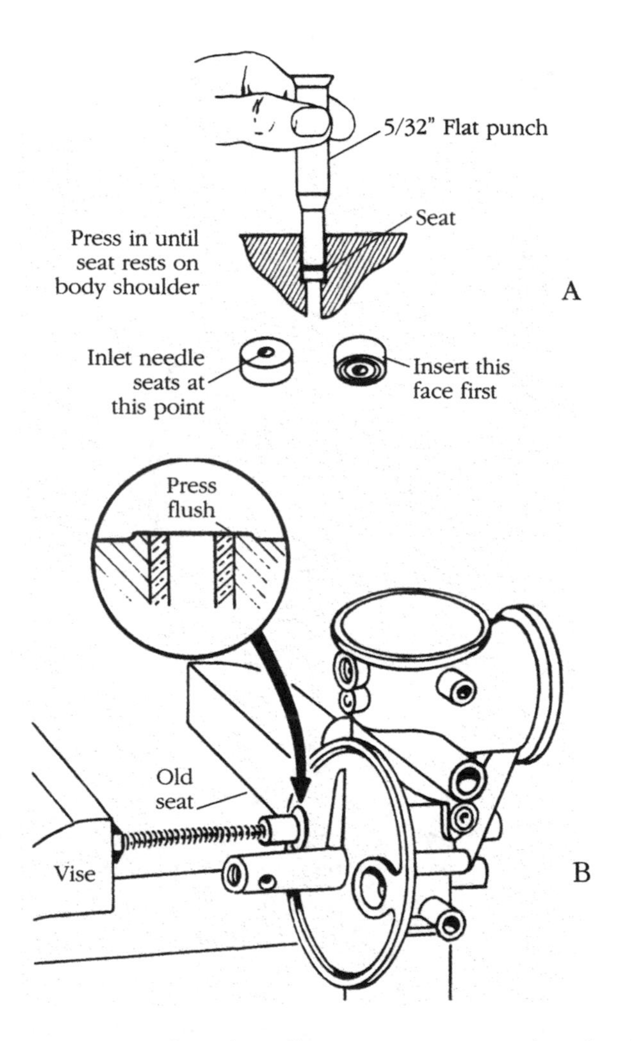

FIGURE 4-10. *Tecumseh and Walbro seats are retrieved with a hooked wire and tapped home with a punch home (A). A self-threading screw can be used to extract Briggs seats, which are pressed in flush, using the original as a cushion (B).*

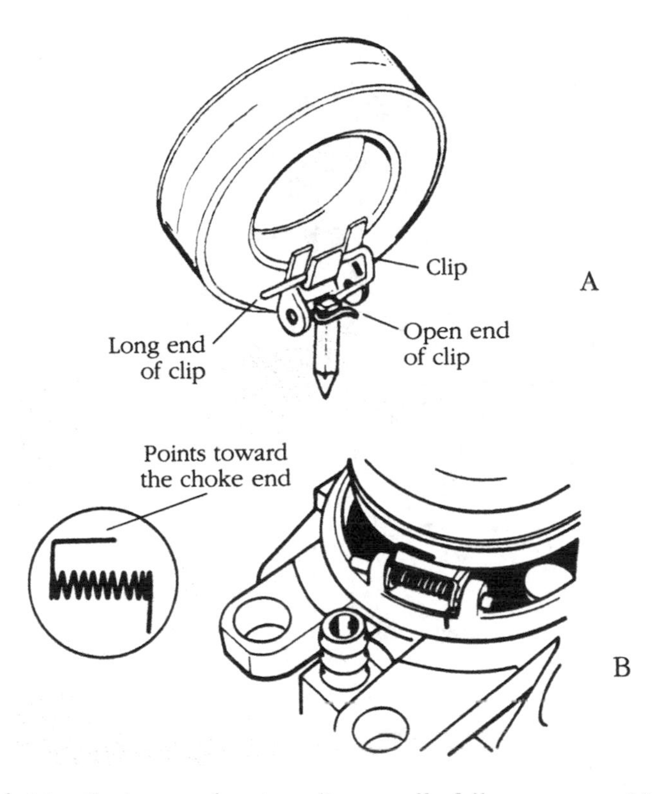

FIGURE 4-11. *Springs and spring clips usually follow an assembly protocol. Drawing A shows one version of an inlet needle clip, which should be installed with the long end of the clip toward the choke. The dampening spring, positioned as shown at B, exerts a slight lift on the float.*

Float adjustments. The position of the float when the inlet valve closes determines the internal fuel level. The higher the float rises before seating the needle, the richer the mixture. Plastic floats install as-is, with no provision for adjustment. Metallic floats, like those shown in Figures 4-12 and 4-13, include tangs that control float height and, usually, float drop. Consult your dealer for the specifications.

Caution: Make the adjustments while holding the float clear of the needle. Forcing the needle into the seat can damage the elastomer tip.

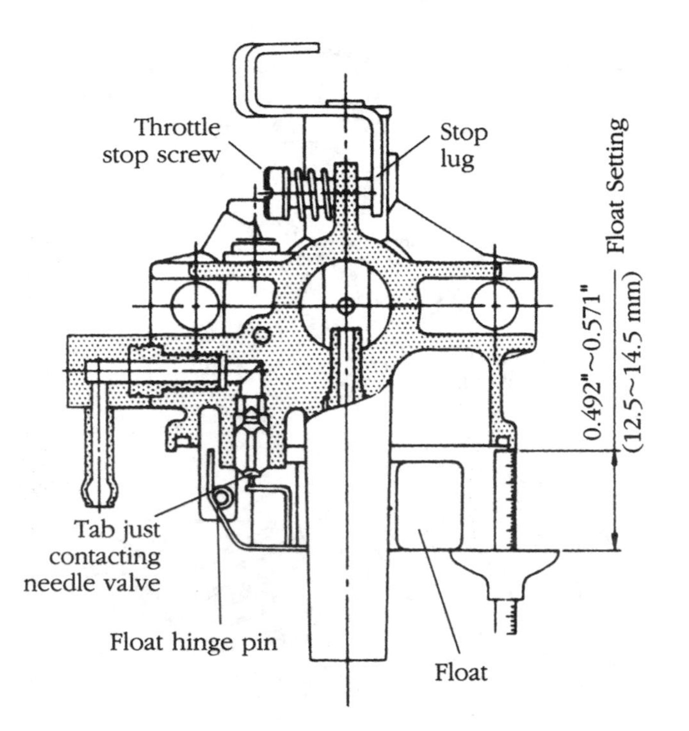

FIGURE 4-12. *This Japanese drawing does a nice job of detailing the inlet seat, float mechanism, and float height that, in this case, is measured from the bottom of the float to the roof of the chamber. Other manufacturers would have you measure from the upper surface of the float.*

Walbro nozzles. The venerable LMG and its LMB cousin supply the idle circuit through a tiny hole in the nozzle drilled after installation. Once the nozzle has been disturbed, the hole no longer indexes. Standard practice is to replace the original nozzle with the LMG or LMB-182 service nozzle, shown in Figure 4-14 and recognized by its annular groove.

It is possible to save the $5 that the service nozzle costs by extracting the lowermost of the two small brass cups on the fuel pickup pedestal. The tang end of a small file ground flat makes the appropriate tool. Set the plug carefully aside and screw in the original nozzle to within an eighth of a turn from fully seated. Gently insert a fine wire (as from a wire brush) into the cup boss while slowly turning the nozzle in and out. You will be able to sense when the wire enters the port. Withdraw the wire and carefully tap the brass plug

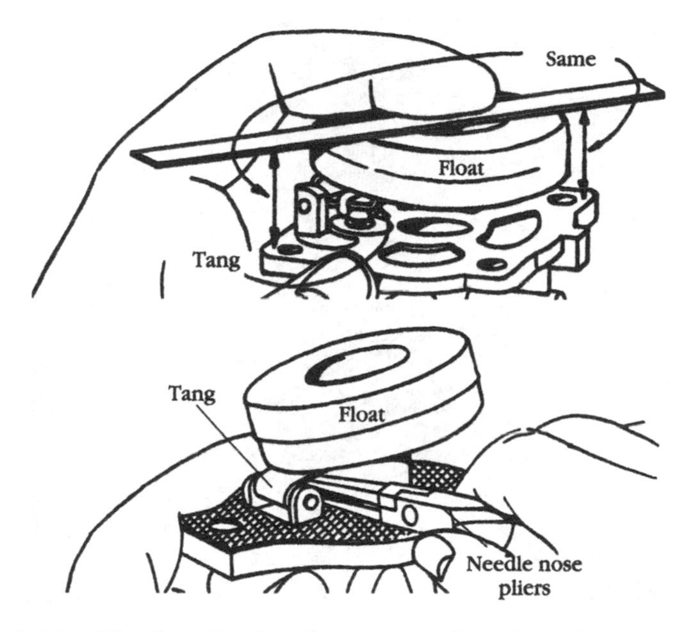

FIGURE 4-13. *The float-height adjustment is best checked with the carburetor inverted. The specification for the unit shown calls for the float to be parallel with the roof of the float chamber.*

home. Thus installed, the original nozzle seems to give better idle performance than the replacement part.

Throttle shaft & bushings. Old-line, fully repairable carburetors, like the Onan in Figure 4-15, support the throttle shaft on replaceable bushings. Wear on these parts creates vacuum leaks that upset the idle calibration and ingest dust.

Remove the throttle butterfly, noting which side is up and outboard, and the throttle shaft. Extract the bushings with a tap or an EZ-Out, and press new bushings in. Reaming should not be necessary. Install the replacement throttle shaft, securing the screws with red Locktite.

Castings. Because the gasket is thick and resilient, over-tightening carburetor mounting screws warps the flange. The gasket surface can be restored with a sheet of medium-grit emery cloth taped to a piece of plate glass or a machine worktable. Apply force to the center of the casting and grind until uniformly bright.

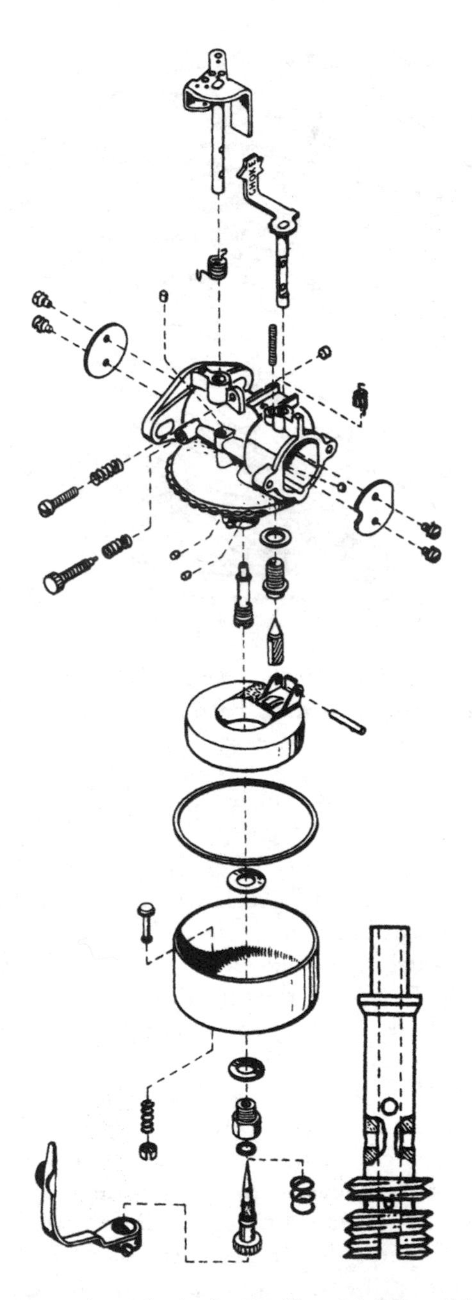

FIGURE 4-14. *Once disturbed, the Walbro LMG and LMB nozzles should be re-aligned as described in the text or replaced with the service nozzle shown on the right.*

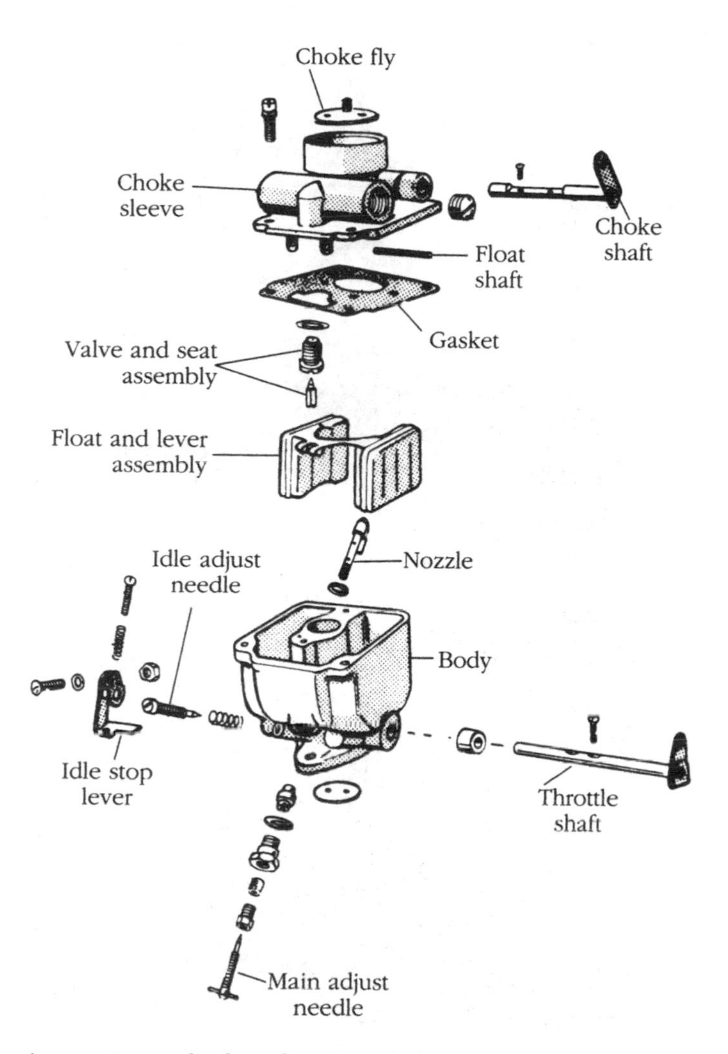

FIGURE 4-15. *Downdraft carburetor used on Onan CCK/CCKA engines. This carburetor is structurally similar to the updraft Briggs Flo-Jet and to the Zenith used on vintage Kohlers. The adjustable main-jet needle threads into the nozzle from below. Throttle shafts and shaft bushings are replaceable.*

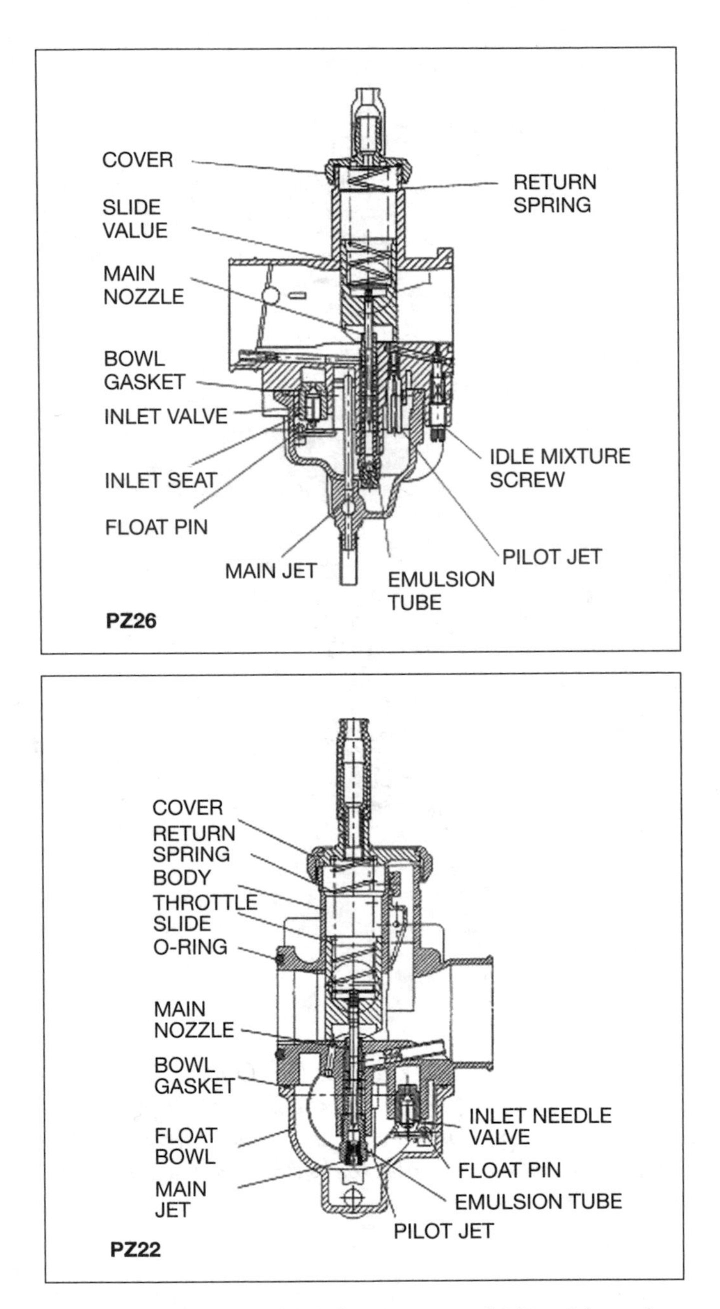

FIGURE 4-16. *Walbro throttle-slide PZ 22 and PZ 26 have become standards for kart racing. Model numbers indicate bore sizes in millimeters.* Walbro Corp.

Slide throttles. Racing and other high-performance carburetors often feature slide throttles that function as variable venturis at low speeds and retract clear of the bore when opened fully (Fig. 4-16). Flow through the high-speed jet is controlled by a tapered needle that moves with the slide. As the throttle opens, the needle lifts to increase the effective size of the jet orifice.

The needles for Walbro PZ and other slide-throttle carburetors can be lowered and raised relative to the slide. The chart in Figure 4-17, showing the relationship between needle position and performance is specific to the PZ, but has application for Mikuni, Dell'Otro and other slide-throttle carburetors.

Diaphragm carburetor operation

Hand-held tools use diaphragm carburetors that operate at any angle, even upside down. And for reasons known only to Tecumseh engineers, the company's mower and edger engines employ the same attitude-tolerant carburetion (Fig. 4-18).

The lower side of the metering diaphragm opens to the atmosphere; the upper, or wetted, side of the diaphragm is exposed to a manifold vacuum. Atmospheric pressure distends the diaphragm upward to unseat the spring-loaded needle. Fuel then fills the cavity above the diaphragm and forces the diaphragm downward to seat the needle. Fuel flow resumes

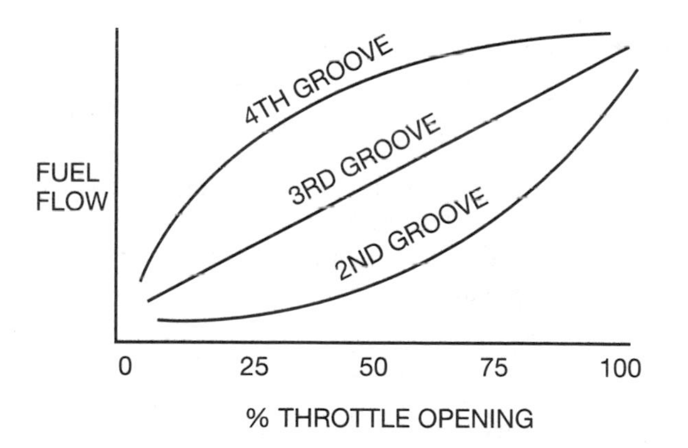

Figure 4-17. *Effects of needle position on flow volume through the main jet at various throttle percentages. Grooves are numbered from the top down. Note that the idle-mixture screw on the PZ22 controls air—backing the screw out admits more air to lean the mixture. The idle-mixture screw on the PZ26 regulates fuel. Backing it out richens the mixture.* Walbro Corp.

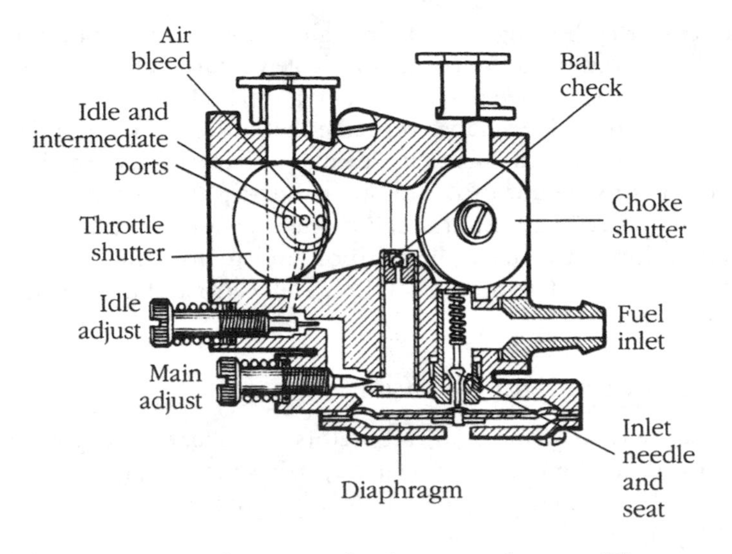

FIGURE 4-18. *Tecumseh-pattern diaphragm carburetor. The cover vent and diaphragm rivet (referenced in text) are clearly shown. Mixture adjustment screws have the same thread but do not interchange. Overtightening the idle-mixture adjust screw can twist its tip off into the jet, where it is almost impossible to extract.*

when the reservoir depletes and the vacuum nudges the diaphragm upward against the needle. In the example shown, the diaphragm bears directly against the needle; other designs transfer diaphragm motion through an adjustable lever.

The unit shown is gravity fed. To operate at any orientation, the carburetor must be supplied by a pump, which nearly always takes the form of a second diaphragm stacked below or alongside the metering diaphragm. You can recognize the pump diaphragm by the two fingerlike cutouts that function as inlet and outlet check valves. When used with two-stroke engines, crankcase pulses activate the diaphragm; four-stroke pumps respond to the changes in inlet-pipe pressure.

Diaphragm carburetors exhibit a variety of special features. The example shown incorporates a ball-check valve in the high-speed circuit to prevent back flow at idle, when the venturi sees little or no vacuum. Many of these carburetors use a butterfly choke valve, as shown. Others employ a primer pump that either squirts raw fuel into the bore or generates air pressure that lifts the diaphragm and unseats the inlet needle. Carburetors for chainsaws and some string trimmers use a cylindrical throttle in lieu of the more familiar butterfly.

When one considers the forces involved—the diaphragm hovers between fuel pressure on one side and atmospheric pressure on the other—it is no surprise that these instruments are temperamental. To complicate matters further, diaphragm carburetors are the focus of intensive engineering efforts aimed at cleaning up the exhausts of two-stroke engines. Some, like over-and-under shotguns, have a second bore to deliver air for scavenging, others incorporate accelerator pumps to reduce the fuel demands on the idle-speed circuit, and several pre-package the fuel delivered by the primer pump to prevent flooding.

If they are honest, few mechanics can claim to be comfortable of this ever-evolving technology, some of which disappears within a year or so of introduction.

Readers who want to delve deeper into these instruments should purchase a Walbro 57-11 tester.[1] The tool detects sticking fuel-pump valves, clogged screens, collapsed hoses, and verifies that the inlet needle pops off and reseats.

Diaphragm carburetor service

Figure 4-19 illustrates potential vulnerabilities for the world's most popular diaphragm carburetor. An exploded view of the same device is shown back at Figure 4-2. Remove the inlet seat with a six-point 9/32-in. socket, ground to fit the seat counterbore. Screwdriver slots milled in the seat generally strip out before the seat budges.

Replace the inlet seat, needle and spring as a matched assembly. Most of these units assemble with the diaphragm gasket on top of the diaphragm; those with an "F" stamped near the air cleaner go together with the diaphragm next to the main casting, followed by the gasket and cover. In all instances, the diaphragm rivet head is up, with the splayed side down toward the vent port on the cover.

Inlet fuel fittings press into the carburetor body and are not disturbed unless the screen, sometimes included in the assembly, is clogged. Twist and pull the fitting out. Press in the replacement to half depth and coat the exposed portion of the shank with Loctite A. Press to depth, with the shoulder up against the carburetor body.

Welch plugs—tiny expansion plugs—must be removed for access to idle and other critical circuits. Some mechanics cut and skate the plugs out with

[1] Walbro Engine Management, Aftermarket Division, 6242-A Garfield St., Cass City, Mich., 48726-1325, (989) 872-2131.

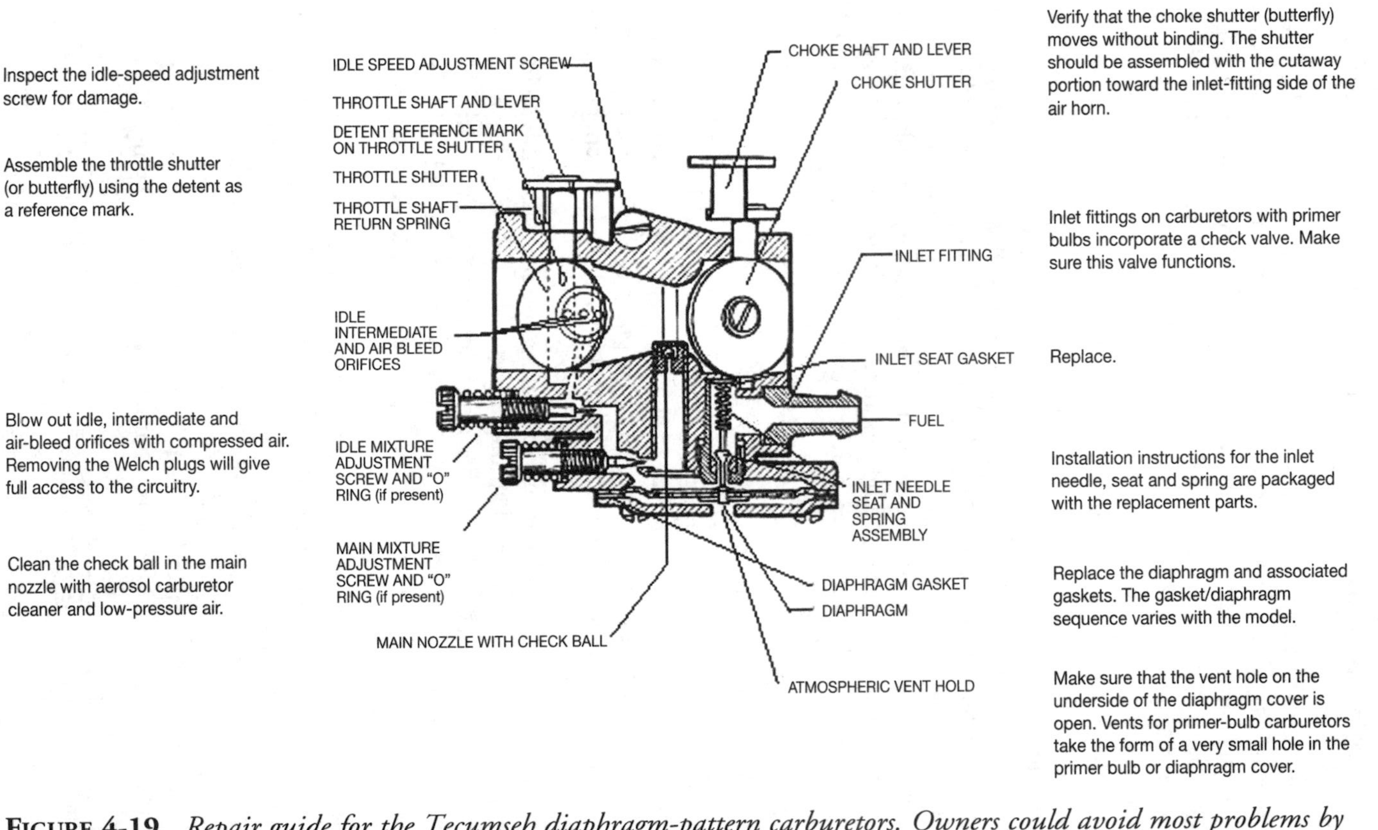

FIGURE 4-19. *Repair guide for the Tecumseh diaphragm-pattern carburetors. Owners could avoid most problems by using fresh, uncontaminated fuel and running the carburetor dry before storage.*

a small cape chisel, others prefer to drill a small hole and extract with an EZ-Out. If you opt for to drill, be sure to blow or vacuum out the swarf. New Welch plugs, packaged in overhaul kits, install with the convex side up. Using a flat punch with the same diameter as plug OD, lightly hammer the plugs home. Seal the edges with nail polish.

The Walbro WYK "pumper" carburetor is a watchmaker's exercise intended for 20–50-cc weed-wacker engines (Fig. 4-20). The WYL incorporates separately both as a throttle and as variable venturi to admit progressively more air as the throttle rotates open. A single jet provides fuel at all speeds. This carburetor also includes a primer bulb (8) that, when depressed, forces air out of the system and, when released, introduces raw fuel into the air horn.

The main metering diaphragm lever is adjusted with a factory gauge as shown in Figure 4-21. Raising the lever opens the inlet needle sooner to richen the mixture. The same effect can be had by substituting a weaker inlet-needle spring for the factory item. Ultra-light aircraft pilots, whose engines verge on meltdown during takeoff, sometimes snip a coil or two off the spring. The surplus fuel helps to cool the pistons. Now that's cutting things close.

Suction-lift carburetor operation

Suction lift carburetors mount on top of the tank and draw fuel through a pickup tube. Several American manufacturers have experimented with the configuration, but modern examples are confined to low-end Briggs & Stratton engines.

Briggs suction-lift carburetors underwent two design evolutions. The earlier version, known as the Vacu-Jet and illustrated back in Figure 4-3, had a single pickup tube and a plug-type choke. As the tank emptied, the fuel level in the pickup tube dropped and the mixture leaned out.

Briggs addressed the problem with the Pulsa-Jet. This carburetor family employs two pickup tubes and an integral pump to move fuel from the main tank to a small reservoir (Fig. 4-22). The carburetor draws from this reservoir, which is continuously topped off by the pump. Thus, the level of fuel in the main tank has no effect upon mixture strength. Depending upon the model, pump diaphragms mount on the outboard side of the carburetor body or between the carburetor and tank (Fig. 4-23).

Suction-lift carburetor service

The most common problem with suction-lift carburetors is failure to deliver fuel caused by a stretched pump diaphragm (Pulsa-Jet) or a sticking check

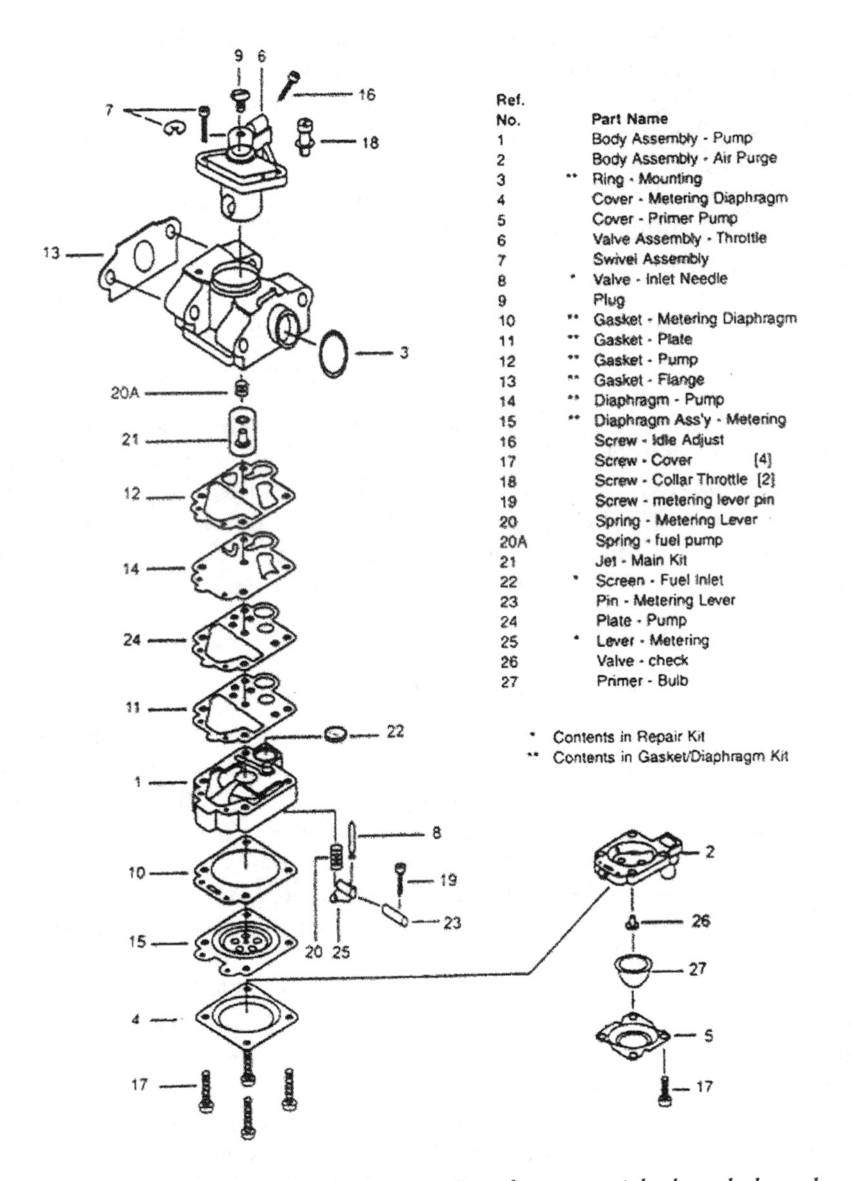

FIGURE 4-20. *Walbro WYK "pumper" carburetor with throttle barrel and metering, pump, and start diaphragms.*

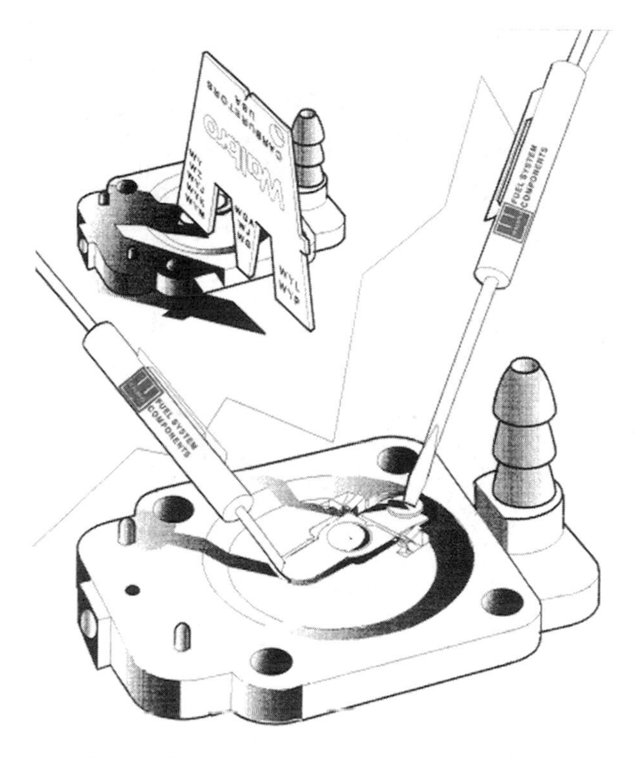

FIGURE 4-21. *Walbro throttle levers adjust with the aid of a factory gauge supplied in tool kit PN 500-500. While holding a small screwdriver against the needle—just hard enough to stabilize it—pass the appropriate gauge over the lever. When adjusted correctly, the lever exerts an almost imperceptible drag on the gauge.*

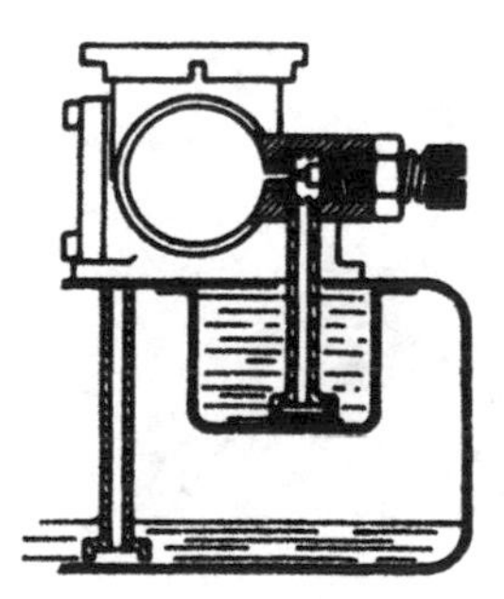

FIGURE 4-22. *Three pulls of the starter cord should pump enough fuel into the Pulsa-Jet reservoir to get the engine started. Once the engine is running, the reservoir fills to the level defined by the spill port.*

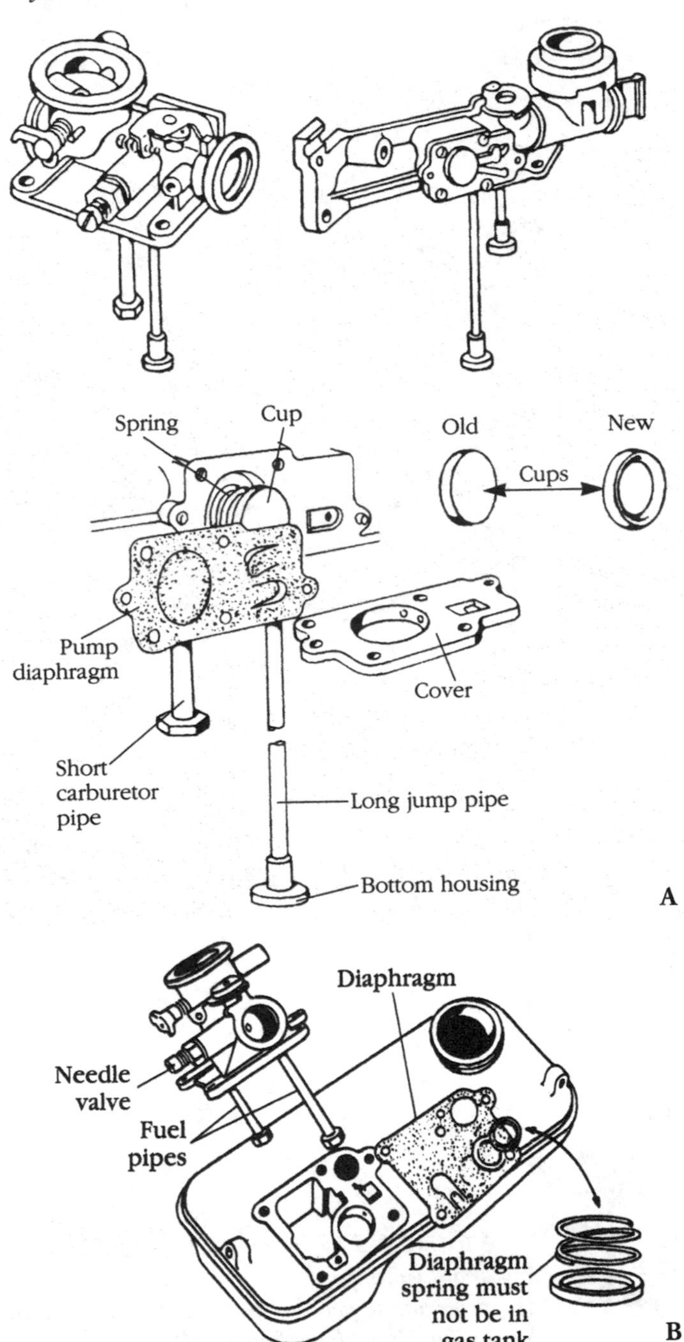

FIGURE 4-23. *Pulsa-Jet malfunctions nearly always involve the diaphragm, which mounts on the side of the carburetor casting (A) or between the carburetor and tank reservoir (B).*

valve (Vacu-Jet). Free the check valve with a piece of thin wire inserted through pickup-tube screen. Low-speed ports on both carburetors can usually be cleared by removing the mixture adjustment screw and applying compressed air to the screw boss.

Pulsa- and Vacu-Jets are sometimes supplied with vacuum-operated chokes. A spring-loaded diaphragm holds the choke closed during cranking. Once the engine starts, a manifold vacuum acts on the underside of the diaphragm to pull the choke open. The choke also functions as an enrichment valve: when the engine slows under load, the loss of manifold vacuum causes the choke to close.

Assembly is a bit tricky:

1. Clip the choke link to the diaphragm as shown in Figure 4-24.
2. Invert the assembly and snake the linkage into its recess (Fig. 4-25).
3. Turn the parts over and tighten the mounting screws just enough for purchase.
4. With one finger holding the choke butterfly closed, attach the actuating link to the choke (Fig. 4-26).
5. Tighten the mounting screws in an X-pattern. Diaphragm preload should hold the choke lightly closed until the engine starts.

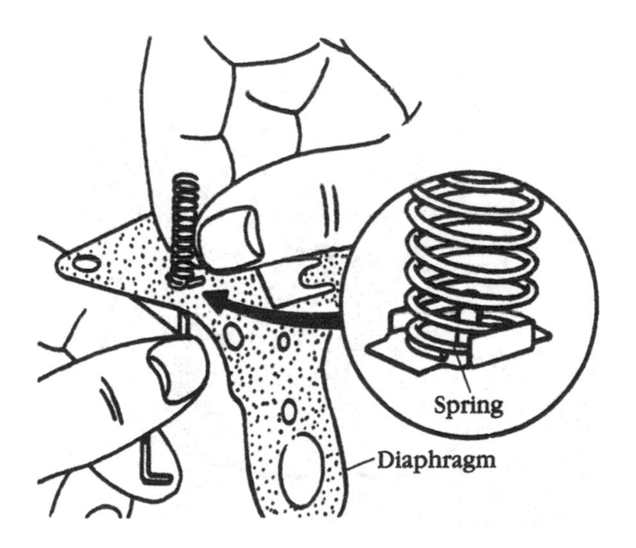

FIGURE 4-24. *The choke link clips to the diaphragm.*

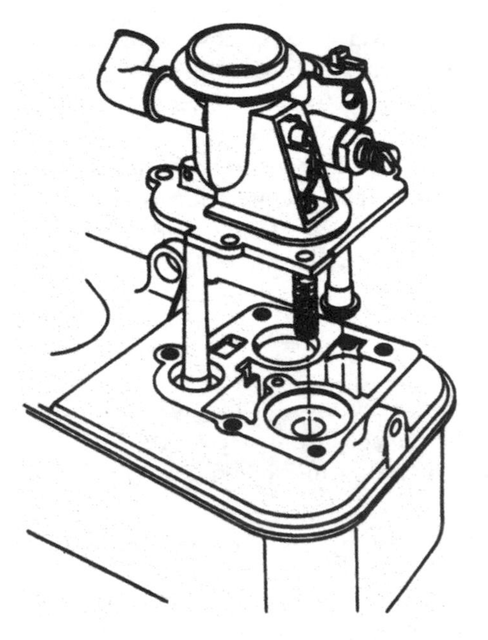

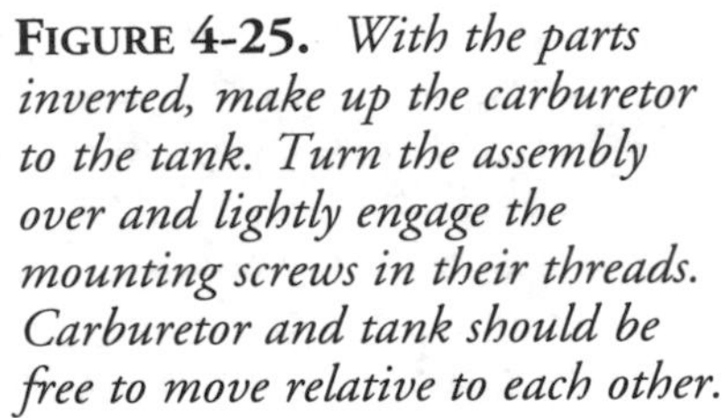

FIGURE 4-25. *With the parts inverted, make up the carburetor to the tank. Turn the assembly over and lightly engage the mounting screws in their threads. Carburetor and tank should be free to move relative to each other.*

FIGURE 4-26. *To establish preload it is necessary to make up the linkage while holding the butterfly closed. When installed correctly, the choke closes when the engine stops, flutters under acceleration and load, and opens at steady speed.*

External adjustments

The classic carburetor has three adjustment screws: idle rpm, idle mixture, and high-speed mixture. The idle-rpm screw functions as a throttle stop and has no effect upon mixture strength. The idle mixture screw threads into the carburetor body near the throttle blade. Backing the screw out richens the mixture. The high-speed screw, variously called the "main adjust needle" or "main fuel needle," works in conjunction with the main jet to regulate mixture strength at wide throttle angles. It may be located under the float bowl or on the carburetor body, upstream of the idle-mixture screw. At midthrottle, there is some interaction between high-speed and idle-mixture adjustments.

Carburetors that comply with federal and California emissions standards curtail adjustment with limiter caps or do away with adjustment screws entirely. Suction-lift and diaphragm carburetors with cylindrical throttles (such as the WYK illustrated previously) have a single screw that affects mixture strength across the rpm band.

Initial adjustments

Backing out the adjustment screws 1 1/2 or 4 turns from lightly seated should admit enough fuel to start the engine. Run down the screws finger-tight; forcing the issue with a screwdriver can damage both the needles and seats (Fig. 4-27).

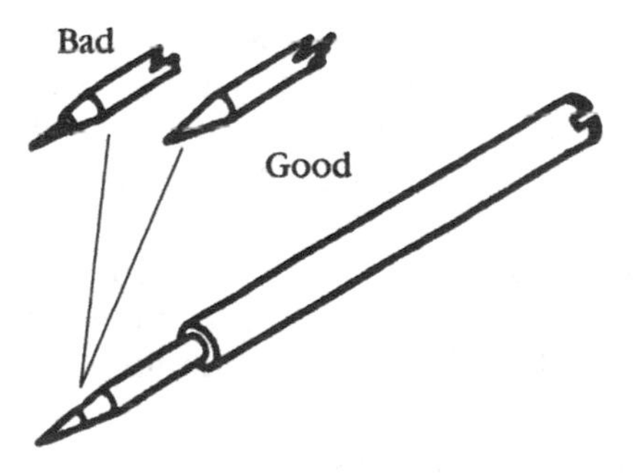

FIGURE 4-27. *Bent needles must be replaced if the carburetor is to be adjusted properly. Wear grooves deep enough to be felt destroy the taper and make adjustment hyper-sensitive.*

Final adjustments

Instructions that follow assume that the carburetor has idle rpm, idle mixture, and high-speed mixture adjustments. If only idle rpm and idle mixture are present, adjust for best idle, richening the mixture if the engine bogs under acceleration or load.

Run the engine until it reaches operating temperature and then, with the throttle about three-quarters open:

1. Back out the high-speed mixture screw in small increments, about an eighth of a turn at a time. Allow a few seconds for the effect of each adjustment to be felt. Stop when engine speed falters at the rich limit and, using the screw slot as a reference, note how far the screw has been turned.
2. Tighten the screw in small increments as before. Stop on the threshold of lean roll, which represents the leanest combustible mixture. Note the position of the screw slot.
3. Split the difference between the onset of lean roll and the rich limit.
4. Close the throttle and adjust the idle mixture screw for the fastest idle. If the engine overspeeds, back off the idle-rpm screw a few turns.
5. Snap the throttle open with your finger. Hesitation can be, up to a point, compensated for by a richer mixture. Many carburetors respond to a slightly richer high-speed mixture; others accelerate more smoothly if the idle-speed mixture is richened. Experiment until you find the formula.
6. Test the engine under load. The main jet may need to pass more fuel at the expense of a slightly rich idle.

Air cleaners

Most four-cycle engines employ pleated paper filters that, when working properly, should trap more than 99 percent of airborne dust particles as small as 10 microns in diameter (1 micron = one millionth of a millimeter or four one-hundred thousandths of an inch.). Figure 4-28 illustrates a paper filter used on Wisconsin engines.

Other than periodic replacement, paper filters require no service. Inspection merely introduces dust into the carburetor air horn, and attempts to clean the filter by knocking the dust loose or blowing it out with compressed air compromise its effectiveness.

Polyurethane elements filter less efficiently than paper, but cost almost nothing to maintain. Clean the element in hot water and detergent. Pour a

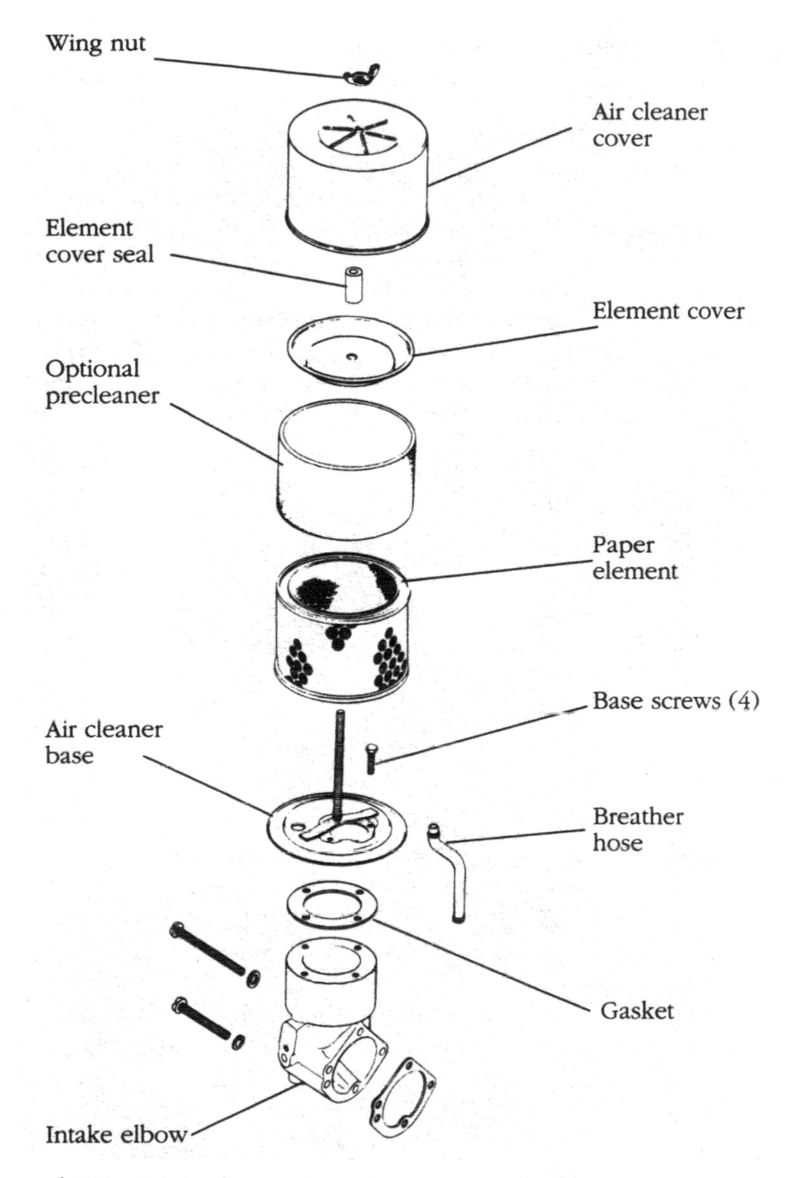

FIGURE 4-28. *Many four-cycle engines come with a two-stage air cleaner, consisting of a polyurethane precleaner and a paper inner element. When installing a new precleaner, turn it inside out to make a good seal with the inner element. The assembly pictured is used on several Wisconsin engines.*

few ccs of motor oil over the sponge, kneading it in gently. For best results, the filter should be reoiled every few hours of operation. Two-stroke engines use polyurethane or cotton gauze filters that are continuously oiled by the fuel mist that hovers around the carburetor air horn.

Fuel pumps

Mechanical pumps drive off a camshaft or crankshaft eccentric (Fig. 4-29) and use interchangeable check valves. Repair procedures are obvious, but a few points deserve mention. Pumps will be damaged if the cover screws are tightened against a taut diaphragm. Move the lever to its travel limit and hold it in that position while drawing down the screws. Apply some grease to the end of the lever where it contacts the cam. Finally, crank the engine over so that the pump mates up without having to struggle against spring tension.

Diaphragm pumps operate from crankcase pressure pulses. Most use cutouts in the diaphragm as check valves and some, like the example in Figure 4-30, include a dampening diaphragm to smooth pressure fluctuations. Scribe mark the housings before disassembly and note the position of the gasket relative to the diaphragm.

Governors

A unique feature of small engines is the governor that helps maintain speed under varying loads and limits maximum speed. Air-vane governors sense engine rpm as a function of air pressure and velocity (Fig. 4-31). The vane mounts under the shroud in the cooling air stream and acts to close the throttle butterfly. The governor spring pulls the butterfly in the opposite direction. If speed drops below a certain value, the spring opens the throttle.

Mechanical governors employ centrifugal flyweights as the speed sensor (Fig. 4-32). The flyweights respond to engine speed by moving outward. This movement, translated through a spool and yoke, applies progressively more force to close the throttle as engine speed increases. The governor spring counteracts this force. As speed drops, the force generated by the flyweights diminishes and the spring pulls the butterfly open.

The more elaborate mechanisms incorporate a low-speed adjustment (distinct from the throttle stop screw) and have provision to adjust governor sensitivity, or speed droop. Most allow the maximum rpm limit to be varied by moving the governor arm relative to the shaft. But no universal procedure applies (Fig. 4-33). Governor adjustments should be left to dealer mechanics.

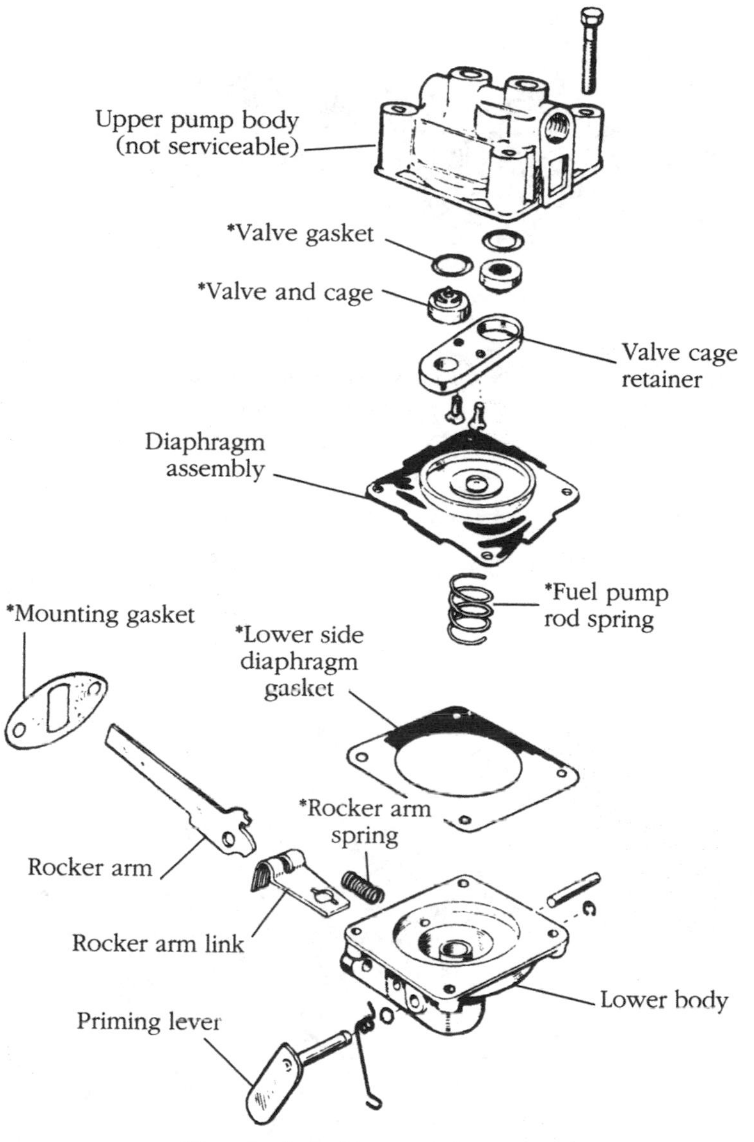

FIGURE 4-29. *Onan mechanical fuel pump employs check valves that can be tested by blowing through them. Valves should block air entering in one direction and pass it in the other.*

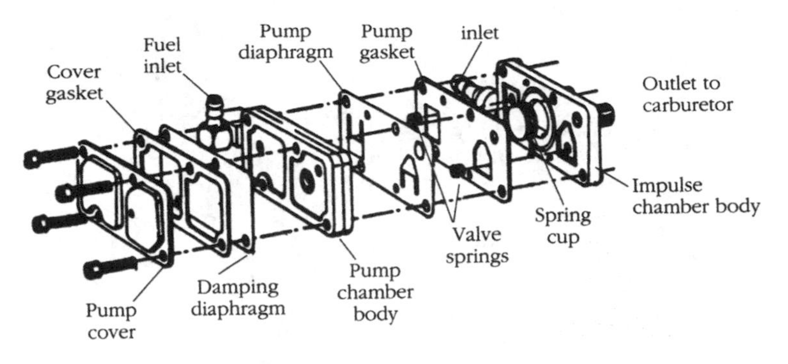

FIGURE 4-30. *Briggs & Stratton diaphragm-type fuel pump includes a second diaphragm to smooth output pressure pulses.*

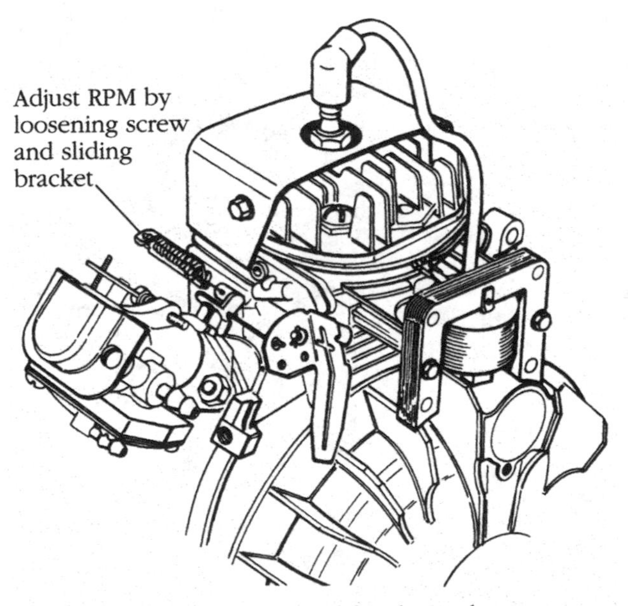

FIGURE 4-31. *Air vane governor used on Tecumseh two-strokes that run at a constant speed determined by the spring-bracket adjustment.*

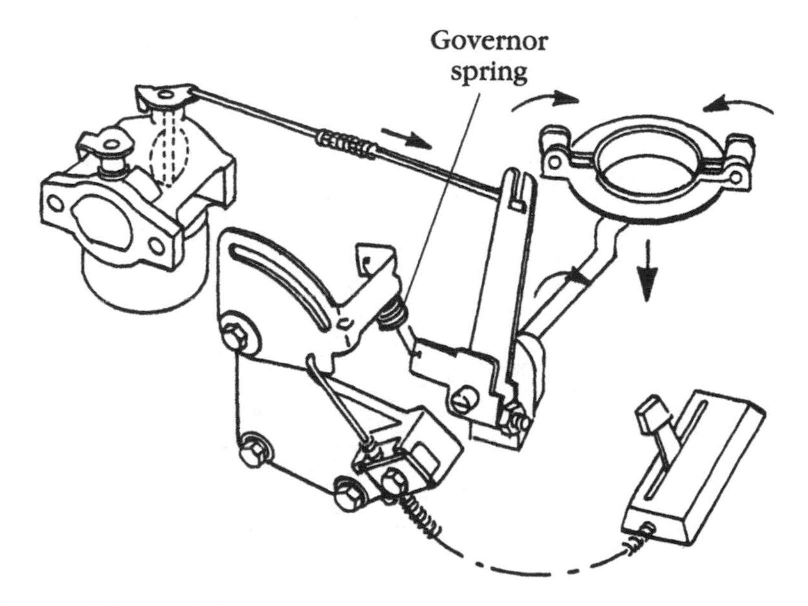

FIGURE 4-32. *Mechanical governors use centrifugal force developed by a pair of spinning weights to close the throttle. The governor spring pulls the throttle open. As the engine slows under load, the force exerted by the weights diminishes and the spring opens the throttle butterfly.*

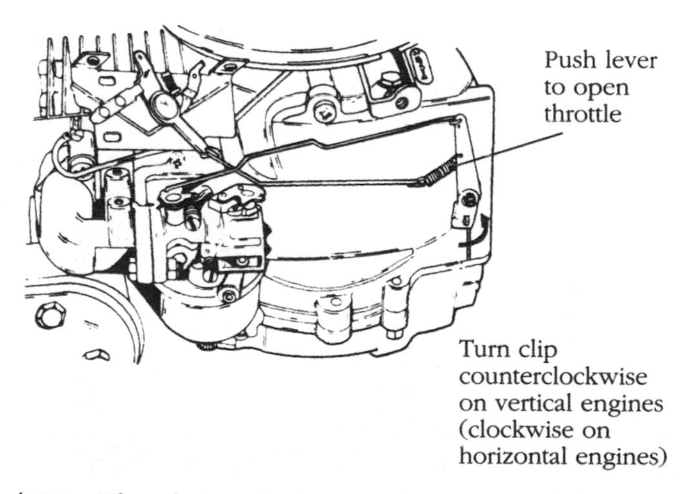

FIGURE 4-33. *The relationship between the governor shaft and wide-open throttle is the crucial aspect of small engine work. Make a mistake here and the engine grenades. But as the illustration demonstrates, procedures are not standardized.*

5

Rewind starters

Unlike other engine systems that operate continuously, manual and electric starters are designed for intermittent use, which is why rewind starters can get by with nylon bushings, and why motor pinions can cheerfully bang into engagement with the flywheel. The starter usually lasts about as long as the engine and the owner is satisfied.

But the balance between starter and engine life goes awry if the engine is allowed to remain chronically out of tune. Most starter failures are the result of overuse: The starter literally works itself to death cranking a bulky engine. Whenever you repair a starter, you must also—if the repair is to be permanent—correct whatever it is that makes the engine reluctant to start in the first place.

Side pull

The side-pull rewind (recoil, self-winding, or retractable) starter was introduced by Jacobsen in 1928 and has changed little in the interim. These basic components are always present:

- Pressed steel or aluminum housing, which contains the starter and positions it relative to the flywheel.
- Recoil spring, one end of which is anchored to the housing, the other to the sheave.
- Nylon starter rope, which is anchored to and wound around the sheave.
- Sheave, or pulley.

- Sheave bushing between sheave and housing or (on vintage Briggs & Stratton) between sheave and crankshaft.
- Clutch assembly.

Troubleshooting

Most failures have painfully obvious causes, but it might be useful to have an idea of what you are getting into before the unit is disassembled.

Broken rope is the most common failure, often the result of putting excessive tension on the rope near the end of its stroke or by pulling the rope at an angle to the housing. The problem is exacerbated by a worn rope bushing (the guide tube, at the point where the rope exits the housing). In general, rope replacement means complete starter disassembly, although some designs allow replacement with the sheave still assembled to the housing.

Refusal of the rope to retract. If the whole length of the rope extends out of the housing, either the spring has broken or the anchored end slipped. If the rope retracts part of the way and leaves the handle dangling, the problem is loss of spring preload. The best recourse is to replace the spring, although preload tension can be increased by one sheave revolution. When this malfunction occurs on a recently repaired unit, check starter housing /flywheel alignment, spring preload tension, and replacement rope length and diameter.

Failure to engage the flywheel is a clutch problem, caused by a worn or distorted brake spring, a loose retainer screw, or oil on clutch friction surfaces. While recoil springs and sheave bushings require some lubrication, starter clutch mechanisms must, as a rule, be assembled dry.

Excessive drag on the rope often results from misalignment between the starter assembly and flywheel. If repositioning the starter does not help, remove the unit, turn the engine over by hand to verify that it is free, and check starter action. The problem might involve a dry sheave bushing.

Noise from the starter as the engine runs should prompt you to check the starter housing and flywheel alignment. On Briggs & Stratton in-house designs, the problem is often caused by a dry sheave bushing (located between the starter clutch and crankshaft). Remove the blower housing and apply a few drips of oil to the crankshaft end.

Overview of service procedures

Rewind starters are special technology, and an overall view of the subject is helpful. The first order of business is to release spring preload tension, which can be done in two ways. Any rewind starter can be disarmed by removing the rope handle and allowing the sheave to unwind in a controlled fashion. Other starters have provisions for tension release with the handle still attached to the rope. Briggs & Stratton provides clearance between sheave diameter and housing that allows several inches of rope to be fished out of the sheave groove. This increases the effective length of the rope, enabling the sheave and attached spring to unwind. Other designs incorporate a notch in the sheave for the same purpose (Fig. 5-1).

Brake the sheave with your thumbs as it unwinds. Count sheave rotations from the point of full rope retraction so that the same preload can be applied upon assembly.

The sheave is secured at its edges by crimped tabs and located by the crankshaft extension (Briggs & Stratton side pull), or else it rotates on a pin attached to the starter housing. A screw (Eaton) or retainer ring (Fairbanks-Morse and several foreign makes) secures the sheave to the post.

The mainspring lives under the sheave, coiled between sheave and housing; with its inner, or movable, end secured to the sheave hub. The outer, or stationary, spring end anchors to the housing. Unless the spring is broken, do not disturb it.

Warning: Even after preload tension is dissipated, rewind springs store energy that can erupt when the sheave is disengaged from the housing. Wear safety glasses.

The manner in which recoil springs secure to the housing varies among makes, and this affects service procedures. Many use an integral spring retainer that indexes to slots in the housing (Fig. 5-2). The spring and retainer are serviced as a unit and should not be separated.

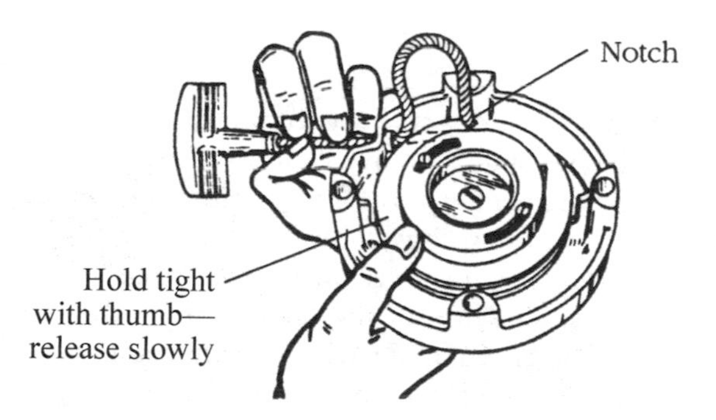

FIGURE 5-1. *Common sense dictates that the starter should be disarmed before the sheave is detached. Most have provision to unwind the rope a turn or so while others are disarmed by removing the rope handle and allowing the rope to fully retract.*

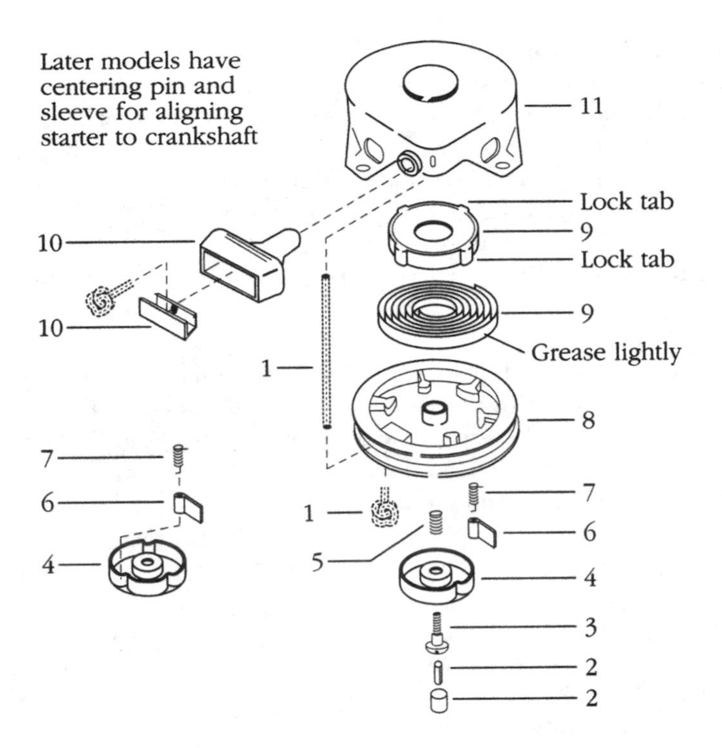

FIGURE 5-2. *Eaton rewind starter, with integral mainspring and housing, should not be dismantled in the field. Lock tabs on the spring-housing rim mean that the spring and housing should not be dismantled. The starter also uses a small coil spring—shown directly below the sheave—to generate friction on the clutch assembly.*

Another attachment strategy is to secure the spring to a post pressed into the underside of the housing. The fixed end of the spring forms an eyelet or hook that slips over the anchor post. To simplify assembly, most manufacturers supply replacement springs coiled in a retainer clip. The mechanic positions the spring and retainer over the housing cavity with the spring eyelet aligned to the post and presses the spring out of the retainer, which is then discarded. Sheave engagement usually takes care of itself. Exceptions are discussed in sections dealing with specific starters.

Some starters adapt to left- or right-hand rotation by reversing the spring (Fig. 5-3). Viewing the starter housing from the underside and using the movable spring end as reference, clockwise engine rotation requires counterclockwise spring windup. The wrap of the rope provides appropriate sheave rotation.

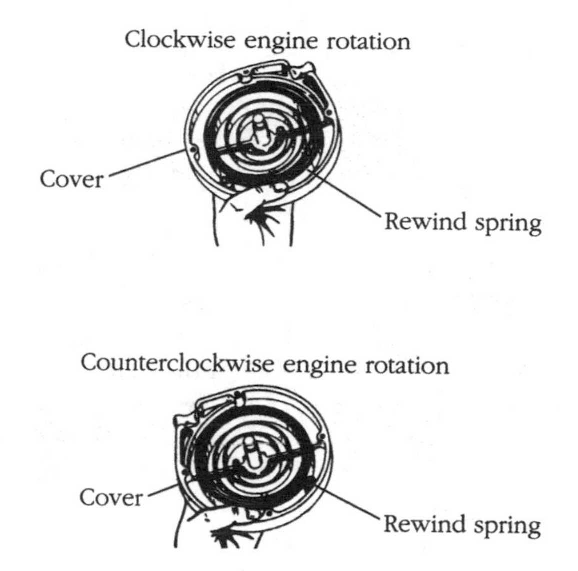

FIGURE 5-3. *Many rewind springs and all ropes can be assembled for left or right hand engine rotation. This feature is a manufacturing convenience that makes life difficult for mechanics.*

The spring anchor for traditional Briggs & Stratton starters takes the form of a slot in the starter housing through which the spring passes. These devices are assembled by winding the spring home with the sheave. Thread the movable end of the spring through the housing slot, engage the movable end with the sheave, and rotate the sheave against the direction of engine rotation until the whole length of the spring snakes through the housing slot. A notch on the end of the spring anchors it to the housing.

Rewind spring preload is necessary to maintain some rope tension when the rope is retracted. Too little preload and the rope handle droops; too much and the spring binds solid.

There are two ways to establish preload. Most manufacturers suggest the following general procedure:

1. Remove the rope handle if it is still attached.
2. Secure one end of the rope to its anchor on the sheave.
3. Wind the rope completely over the sheave, so that the sheave will rotate in the direction of engine rotation when the rope is pulled.
4. Wind the sheave against engine rotation a specified number of turns. If the specification is unknown, wind until the spring coil binds, then release the sheave for one or two revolutions.

5. Without allowing the sheave to unwind further, thread the rope through the guide tube (also called a ferrule, bushing, or eyelet) in the starter housing and attach the handle.
6. Gently pull the starter through to make certain the rope extends to its full length before the onset of coil bind and that the rope retracts smartly.

Another technique can be used when the rope anchors to the inboard (engine) side of the sheave:

1. Assemble sheave and spring.
2. Rotate the sheave, winding the mainspring until coil bind occurs.
3. Release spring tension by one to no more than two sheave revolutions.
4. Block the sheave to hold spring tension. Some designs have provisions for a nail that is inserted to lock the sheave to the housing; others can be snubbed with Vise-Grips or C-clamps.
5. With rope handle attached, thread rope through housing ferrule and anchor it to the sheave.
6. Release the sheave block and, using your thumbs for a brake, allow the sheave to rewind, pulling rope after it.
7. Test starter operation.

The starter rope should be the same weave, diameter, and length as the original. If the required length is unknown, fix the rope to the sheave, wind the sheave until coil bind—an operation that also winds the rope on sheave—and then allow the sheave to back off for one or two turns. Cut the rope, leaving enough surplus for handle attachment.

Three types of clutch assemblies are encountered: Briggs & Stratton sprag, or ratchet; Fairbanks-Morse friction-type; and the positive-engagement dog-type used by other manufacturers. In the event of slippage, clean the Briggs clutch and replace the brake springs on the other types. Fairbanks-Morse clutch dogs respond to sharpening.

One last general observation concerns starter positioning: Whenever a rewind starter has been removed from the engine or has vibrated loose, starter clutch/flywheel hub alignment must be reestablished. Follow this procedure:

1. Attach the starter or starter/blower housing assembly loosely to the engine.
2. Pull the starter handle out about 8 in. to engage the clutch.
3. Without releasing the handle, tighten the starter hold-down screws.
4. Cycle the starter a few times to check for possible clutch drag or rope bind. Reposition as necessary.

Briggs & Stratton

Briggs & Stratton side-pull starters are special in several respects (Fig. 5-4). In addition to its basic function of transmitting torque from the starter sheave to the flywheel and disengaging when the engine catches, the starter clutch also serves as the flywheel nut and starter sheave shaft. Starter and blower housing assembles are integral. It is possible, however, to drill out the spot welds and replace the starter assembly as a separate unit. Bend-over tabs locate the starter sheave in the starter housing.

Disassembly

Follow this procedure:

1. Remove blower housing and starter from engine.
2. Remove rope by cutting the knot at the starter sheave (visible from underside of blower housing).
3. Using pliers, grasp the protruding end of the mainspring and pull it out as far as possible (Fig. 5-5). Disengage the spring from the sheave by rotating the spring a quarter turn or by prying one of the tangs up and twisting the sheave.
4. Clean and inspect. Replace the rope if it is oil-soaked or frayed. Although it might appear possible to reform the end of a broken Briggs & Stratton mainspring, such efforts are in vain and the spring must be

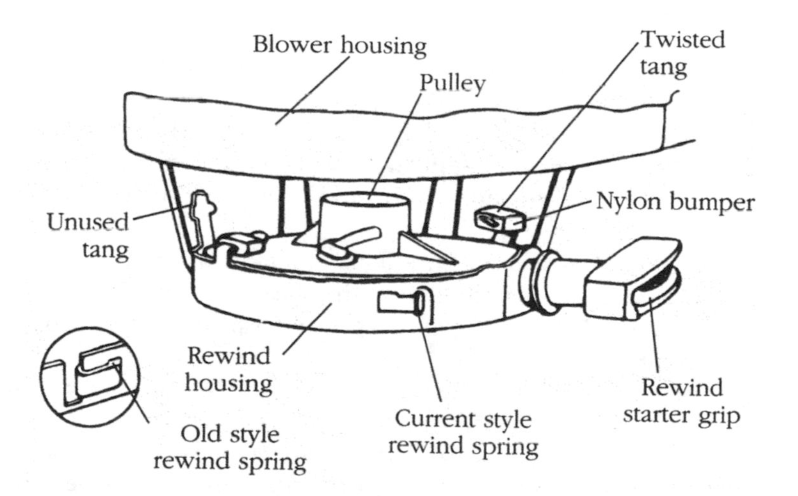

FIGURE 5-4. *Briggs & Stratton rewind starter used widely in the past and carried over today in the "Classic" line.*

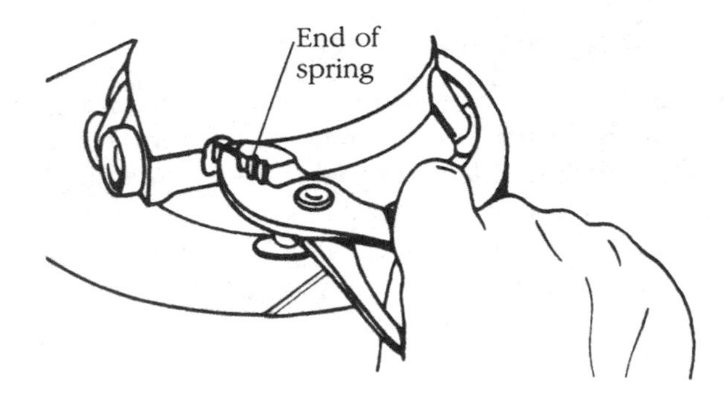

FIGURE 5-5. *Once the rope is removed, pull the rewind spring out of the starter housing. The spring can be detached from the sheave by twisting the sheave a quarter turn.*

replaced for a permanent repair. The same holds for the spring anchor slot in the housing. Once an anchor has swallowed a spring, the housing should be renewed.

Assembly

1. Dab a spot of grease on the underside of the steel sheave. Note that a plastic version requires no lubrication (Fig. 5-6).
2. Secure the blower housing engine-side up to the workbench with nails or C-clamps.
3. Working from the outside of the blower housing, pass the inner end of the mainspring through the housing anchor slot. Engage the inner end with the sheave hub.
4. Some mechanics attach rope (less handle) to the sheave at this point. The rope end is cauterized in an open flame and is knotted.
5. Bend tabs to give the sheave 1/16-in. endplay. Use nylon bushings on models so equipped.
6. Using a 1/4-in. wrench extension bar or a piece of one-by-one inserted into the sheave center hole, wind the sheave 16 turns or so counterclockwise until the full length of the mainspring passes through the housing slot and coil binds.
7. Release enough mainspring tension to align the rope anchor hole in the sheave with the housing eyelet.
8. Temporarily block the sheave to hold spring tension with a Crescent wrench snubbed between the winding tool and the blower housing (Fig. 5-7A).

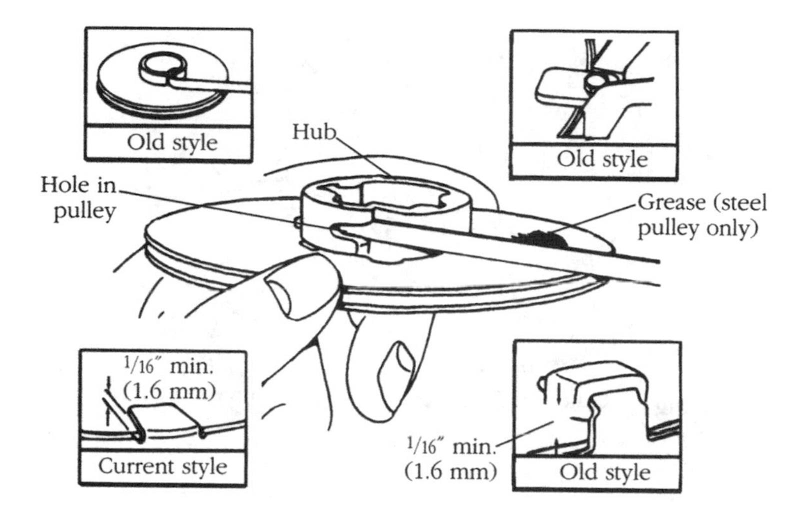

FIGURE 5-6. *Spring installation varies slightly with the date of manufacture. Steel sheaves require lubrication.*

9. If the rope has been installed, extract the end from between the sheave flanges, thread through eyelet, cauterize, and attach the handle. If the rope has not been installed, pass the cauterized end through the eyelet from outside the housing, between sheave flanges, and out through the sheave anchor hole (Fig. 5-7). Knot the end of the rope. Old-style sheaves incorporate a guide lug between flanges. The rope must pass between the lug and sheave hub. This operation is aided by a small screwdriver or a length of piano wire (Fig. 5-7A).

The clutch is not normally opened unless wear or accumulated grim causes it to slip. Older assemblies are secured with a wire retainer clip; newer versions depend upon retainer-cover tension and can be pried apart with a small screwdriver (Fig. 5-8). Clean parts with a dry rag (avoid the use of solvent). The clutch housing can be removed from the crankshaft using a special factory wrench described in Chap. 3. Assemble the unit dry, without lubricant.

Eaton

Recognizable by P-shaped engagement dogs, or pawls, Eaton starters have been used widely on American-made engines. Light-duty models employ a

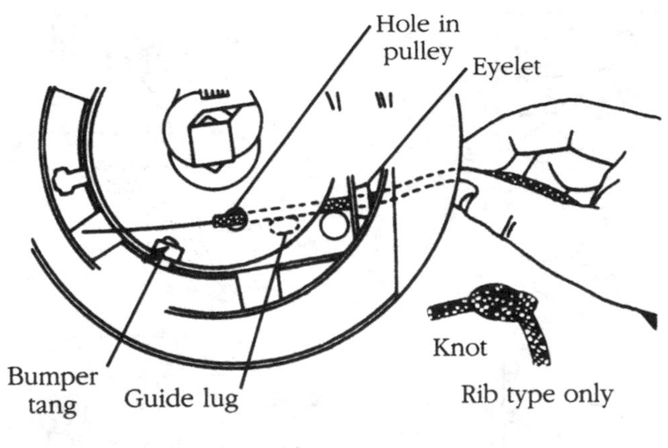

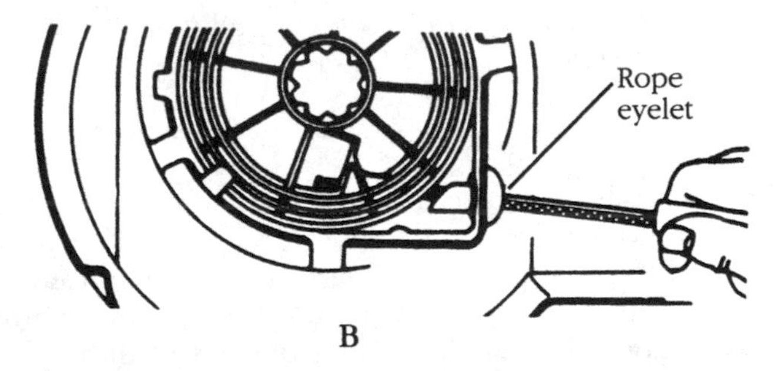

FIGURE 5-7. *Starters for most cast-iron block Briggs engines have an internal rope guide in the form of a lug buried deeply within the sheave. Use a length of piano wire to thread the rope past the inner side of the lug as shown (A). Newer designs omit the guide lug, making installation easier (B).*

single pawl (Fig. 5-9); heavier-duty models use two and sometimes three pawls. All of these starters incorporate a sheave-centering pin, usually riding on a nylon bushing.

A common complaint is failure to engage the flywheel. This difficulty can be traced to the clutch brake, which generates friction that translates into pawl engagement, or to the pawls themselves. Two brake mechanisms are encountered. The later arrangement, shown in Figure 5-9, employs a small coil spring that reacts against the cup-like pawl retainer.

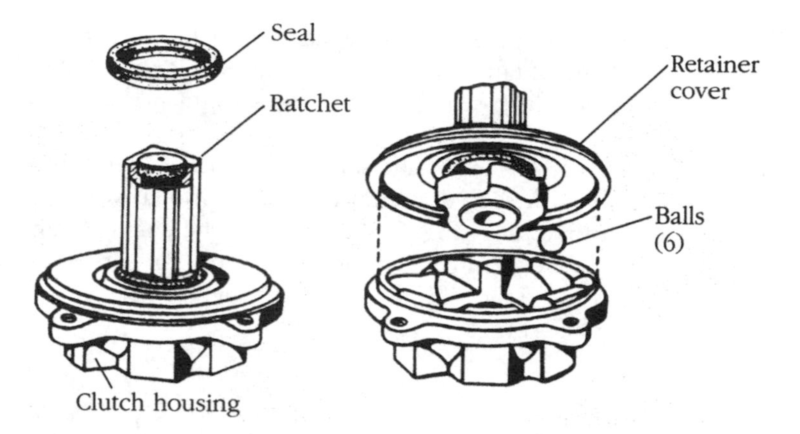

FIGURE 5-8. *Current production clutch cover is a snap fit to the clutch housing. Older versions employed a spring wire retainer. As a point of interest, older engines can be modified to accept new clutch assembly by trimming 3/8 in., from the crankshaft stub and 1/2 in. from the sheave hub.*

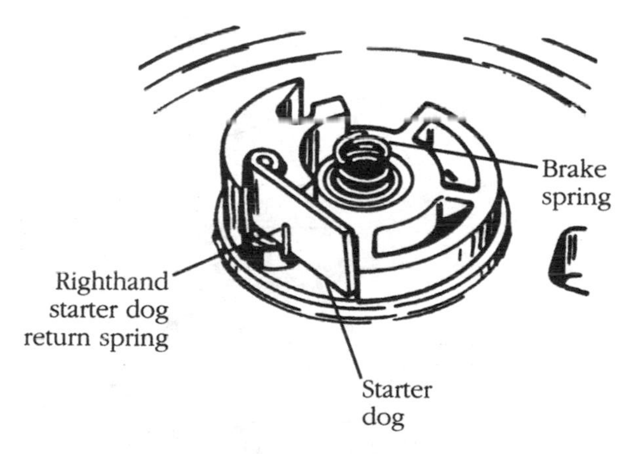

FIGURE 5-9. *Eaton rewind starter partially disassembled. Generous retainer-screw torque compresses brake spring, generating friction against the retainer that causes it to extend the dog. Because the rope attaches to the engine—and accessible—side of the sheave, the rope can be replaced by applying and holding mainspring pretension. The original rope is fished out, new rope is passed through the eyelet and sheave hole, knotted, and pretension is slowly released. As the spring uncoils, it winds rope over sheave.*

The earlier brake interposes a star-shaped washer between the pawl retainer and brake spring. Figures 5-10 and 5-11 show this part. A shouldered retainer screw secures the assembly to the sheave and preloads the brake spring (Fig. 5-12).

Check the retainer screw, which should be just short of "hernia tight"; inspect friction parts, with special attention to the optional star brake; and check the pawl return spring (Fig. 5-9), which can be damaged by engine kickbacks. Clean parts, assemble without lubricant, and observe the response of the pawls as the rope is pulled. If necessary, replace the star brake, retainer cup, and brake spring.

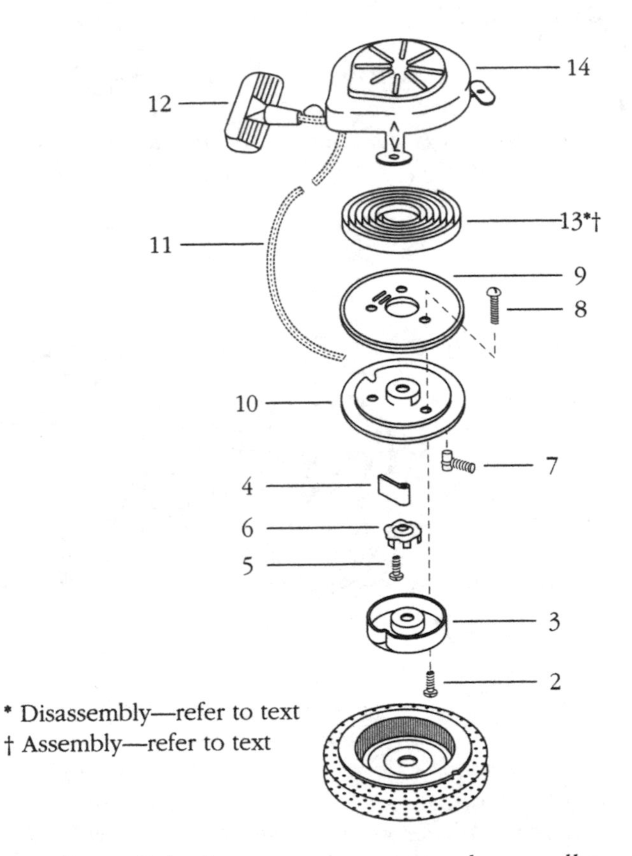

FIGURE 5-10. *Eaton light-duty pattern starter used on small two- and four-stroke engines. This starter is distinguished by its uncased mainspring (13) and single-dog clutch (dog shown at 4, clutch retainer at 3). In the event of slippage during cranking, replace friction spring 5 and brake 6.*

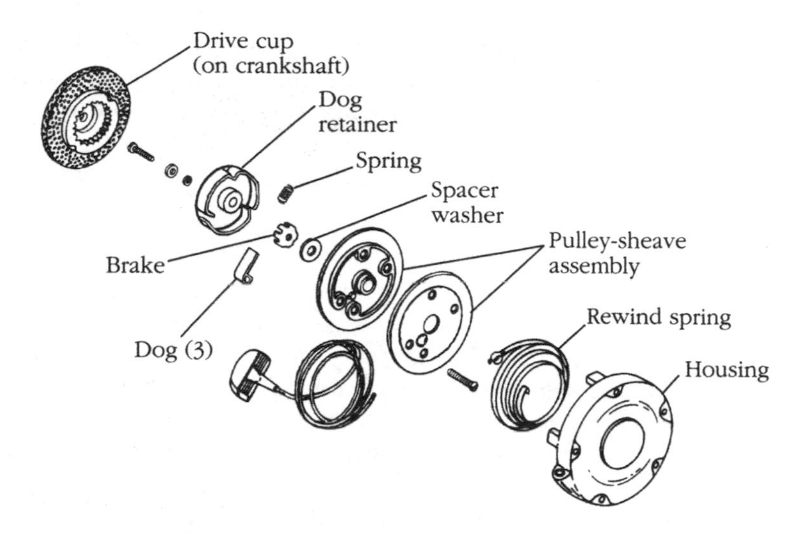

FIGURE 5-11. *Eaton heavy-duty starter—the type used on some Kohler engines. Note the brake friction washer, three-dog clutch, and split sheave.*

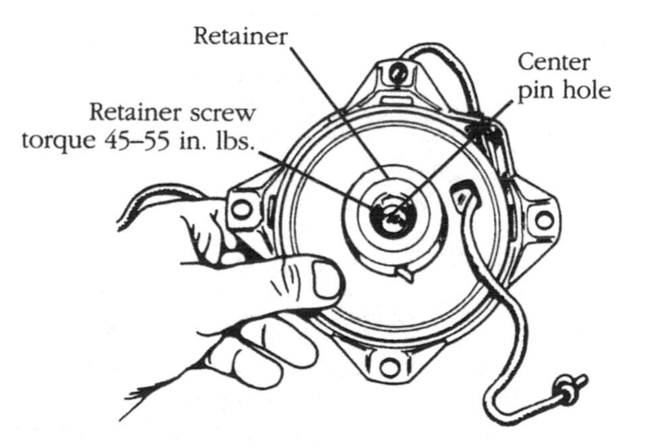

FIGURE 5-12. *View of the engine side of the sheave with a single-dog clutch. This unit is to be assembled dry; only snow proof models, distinguished by a half-moon cam that engages the dogs, require oil on dog-mounting posts.*

Figure 5-11 shows the top-of-the-line Eaton starter used on industrial engines. Service procedures are slightly more complex than for lighter-duty units because the sheave is split. This makes rope replacement more difficult, and the mainspring, which is not held captive in a retainer, can thrash about when the sheave is removed.

Disassembly

1. Remove the five screws securing the starter assembly to the blower housing.
2. Release spring preload. Most heavy-duty models employ a notched sheave that allows rope slack for disarming (see Fig. 5-1).
3. Remove the retainer screw and any washers that might be present.
4. Lift off the clutch assembly, together with the brake spring and the optional brake spring washer.
5. Carefully extract the sheave, keeping the mainspring confined within the starter housing.
   Warning: Wear safety glasses during this and subsequent operations.
6. Remove the rope, which may be knotted on the inboard side of the sheave or sandwiched between sheave halves as shown in Figure 5-11. The screws that hold the sheave halves together can require a hammer impact tool to loosen.
7. Remove the spring if it is to be replaced. Springs without a retainer are unwound one coil at a time from the center outward.
8. Clean and inspect with particular attention to the clutch mechanism. Older light-duty and medium-duty models employed a shouldered clutch retainer screw with a 10-32 thread. This part can be updated to a 12-28 thread (Tecumseh PN. 590409A) by retapping the sheave pivot shaft.

Assembly

1. Apply a light film of grease to the mainspring and sheave pivot shaft. Do not over lubricate because the brake spring and clutch assembly must be dry to develop engagement friction. Snowproof clutches, recognizable by application and by their half-moon pawl cam, might benefit from a few drops of oil on the pawl posts.
2. Install the rewind spring. Loose springs are supplied in a disposable retainer clip. Position the spring—observing correct engine rotation as shown in Figure 5-3—over the housing anchor pin. Gently cut the tape holding the spring to the retainer, retrieving the tape in segments. Install spring and retainer sets by simply dropping them in place.

3. Install the rope, an operation that varies with sheave construction:
   Split sheave
   A. Double-knot the rope, cauterize, and install between sheave halves, trapping the rope in the cavity provided.
   B. Install the sheave on the sheave pivot shaft, engaging the inner end of the mainspring. A punch or piece of wire can be used to snag the spring end as shown in Figure 5-13. Install the clutch assembly.
   C. Wind the sheave until the mainspring coil binds (Fig. 5-14).
   D. Carefully release spring tension two revolutions and align the rope end with the eyelet in the starter housing.
   E. Using Vise-Grips, clamp the sheave to hold spring tension and guide the rope through the eyelet. Attach the handle.
   F. Verify that sufficient pretension is present to retract rope.

   One-piece sheave
   A. Wind the sheave to coil bind and back off to align the rope hole on the inboard face of the sheave with the housing eyelet.
   B. Clamp the sheave.
   C. Cauterize the ends of the rope and install the rope through the eyelet and sheave (Fig. 5-15).

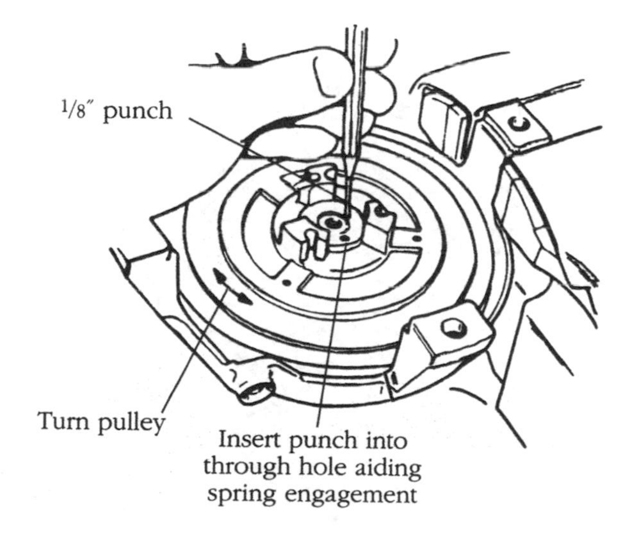

FIGURE 5-13. *A punch aids sprint-to-sheave engagement on large Eaton starters.*

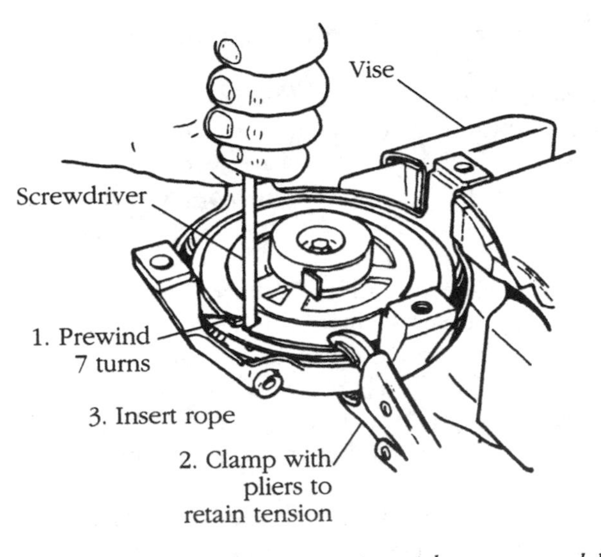

FIGURE 5-14. *Prewind specification varies with starter model and mainspring condition.*

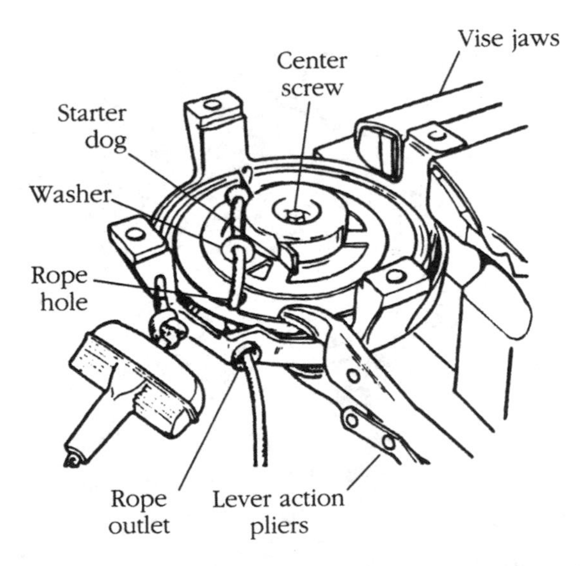

FIGURE 5-15. *Installing the rope on a one-piece sheave involves holding pretension with Vise-Grips and inserting the rope from outside the starter housing, through the eyelet, and into its anchor.*

   D. Knot the rope under the sheave and install the handle.
   E. Carefully release the sheave, allowing the rope to wind as the spring relaxes.
   F. Test for proper pretension.
4. Pull out the centering pin (where fitted) so that it protrudes about 1/8 in. past the end of the clutch retainer screw. Some models employ a centering-pin bushing.
5. Install the starter assembly on the engine, pulling the starter through several revolutions before the hold-down screws are snubbed. Test operation.

Fairbanks-Morse

Fitted to several American engines, Fairbanks-Morse starters can be recognized by the absence of serrations on the flywheel cup. The cup is a soft aluminum casting, and friction shoes (that other manufacturers call "dogs") are sharpened for purchase. Vintage models used a wireline in lieu of the rope. Figure 5-16 is a composite drawing of Models 425 and 475, intended for large single-cylinder engines.

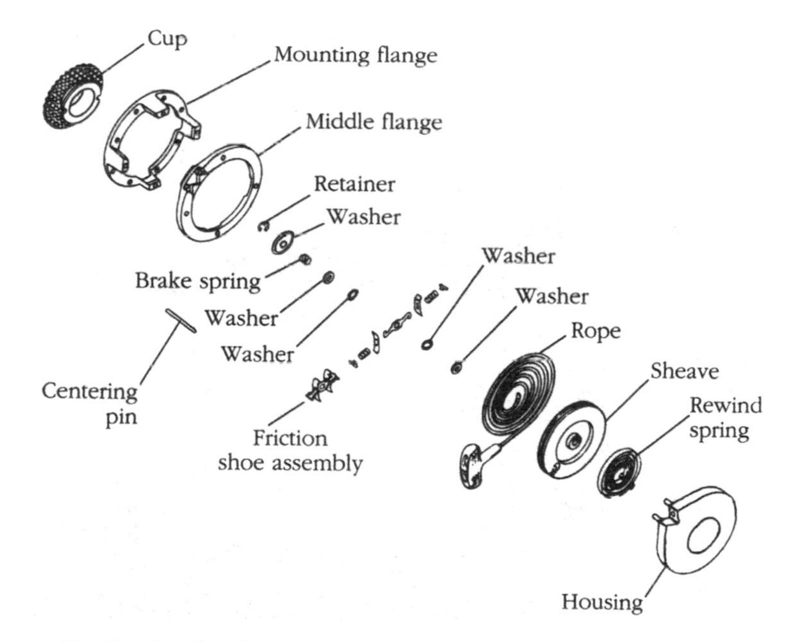

FIGURE 5-16. *Fairbanks-Morse starter used on Kohler and other heavy-duty engines. Mounting and middle flanges are characteristic of Model 475.*

Disassembly

1. Remove the starter assembly from the blower housing.
2. Turn the starter over on bench and, holding the large washer down with thumb pressure, remove the retainer ring that secures the sheave and clutch assembly (Fig. 5-17A).
3. Remove the washer, brake spring, and friction shoe assembly. Normally, the friction shoe assembly is not broken down further.
4. Relieve mainspring preload by removing the rope handle and allowing the sheave to unwind in a controlled fashion. Tension on the Model 475 can be released by removing the screws holding the middle and mounting flanges together (Fig. 5-17B).
5. Cautiously lift the sheave about 1/2 in. out of the housing and detach the inner spring end from the sheave hub.
6. Leave the mainspring undisturbed (unless you are replacing it). To remove the spring, lift one coil at a time, working from the center outward. Wear eye protection.
7. Clean all parts in solvent and inspect.

Assembly

1. Install the spring, hooking the spring eyelet over the anchor pin on the cover. The spring lay shown in Figure 5-17D is for conventional—clockwise when facing flywheel—engine rotation.
2. Rope installation and preload varies with the starter model. In all cases, the rope is attached to the sheave and wound on it before the sheave is fitted to the starter cover and mainspring. The Model 475 employs a split rope guide, or ferrule, consisting of a notch in the middle flange and in the starter housing. Consequently, the rope can be secured to and wound over the flange with the rope handle attached. Model 425 and most other Fairbanks-Morse starters use a one-piece ferrule and the rope must be installed without a handle. After the sheave is secured and the preload established, the rope is threaded through the ferrule for handle attachment.
3. Lubricate the sheave shaft with light grease and apply a small quantity of motor oil to the mainspring. Avoid over lubrication.
4. Install the sheave over the sheave shaft with the rope fully wound. With a screwdriver, hook the inner end of the spring into the sheave hub (Fig. 5-17E).
5. Establish preload—four sheave revolutions against the direction of engine rotation for Model 425, five turns for Model 475, and variable for others.

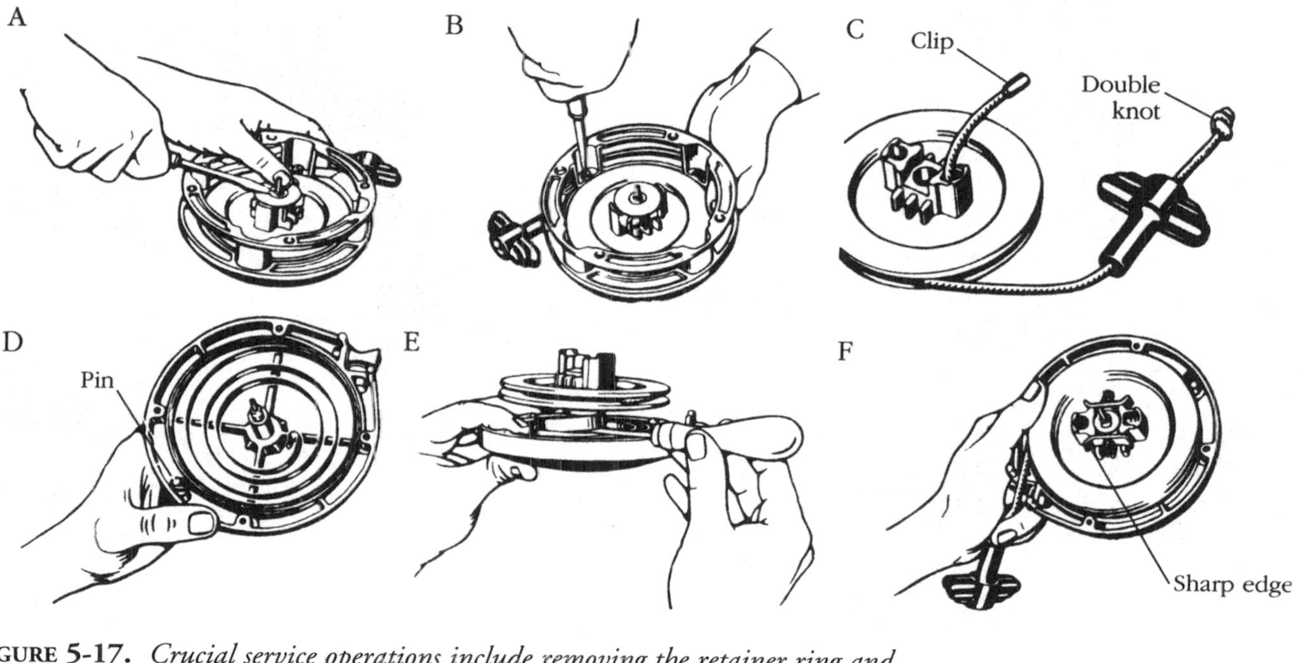

FIGURE 5-17. *Crucial service operations include removing the retainer ring and spring-loaded washer (A), releasing residual spring tension (B), rope anchors and rope lay for standard engine rotation (C), mainspring orientation for standard rotation (D), spring and sheave engagement (E), and correct brake-shoe assembly (F).*

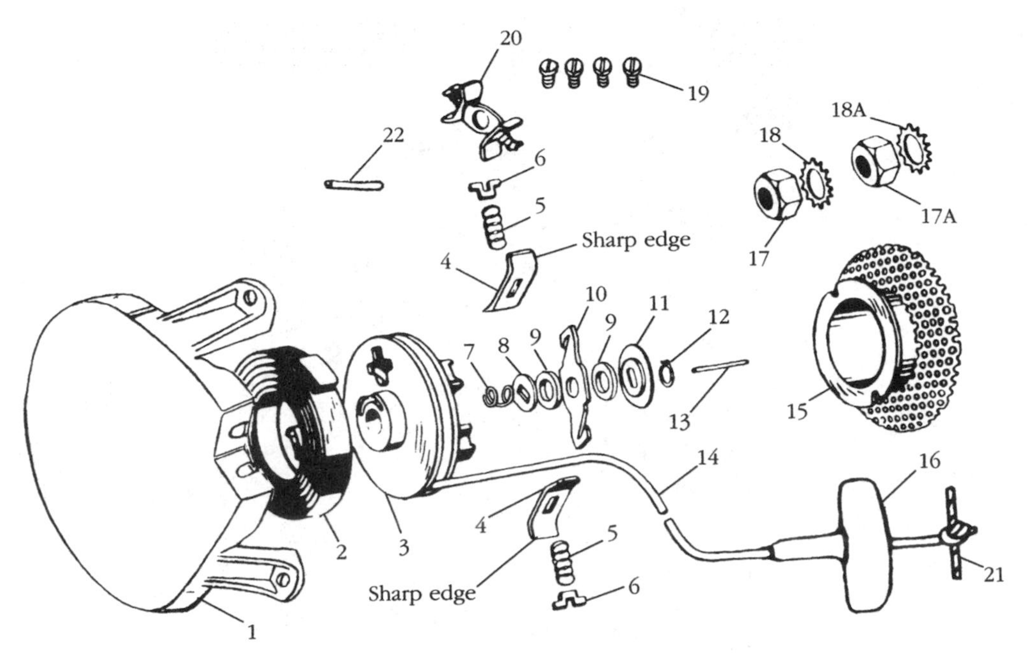

ILLUS. NO.	QTY.	DESCRIPTION
1	1	Cover
2	1	Rewind spring
3	1	Rotor
4	2	Friction shoe plate
5	2	Friction shoe spring
6	2	Spring retainer plate
7	1	Brake spring
8	1	Brake washer
9	2	Fiber washer
10	1	Brake lever
11	1	Brake retainer washer
12	1	Retainer ring
13	1	Centering pin
14	1	Cord
15	1	Cup and screen
16	1	T-handle
17	1	L.H. thick hex nut
17A	1	R.H. thick hex nut
18	1	Ext. tooth lockwasher (left hand)
18A	1	Ext. tooth lockwasher (right hand)
19	4	Pan hd. Screw w/int.-ext. tooth lockwasher
20	1	Friction shoe assembly, includes: Items 4, 5, 6 and 10
21	1	Spiral pin
22	1	Roll pin

FIGURE 5-18. *Small series Fairbanks-Morse can accommodate right- and left-hand engine rotation.*

6. Complete the assembly, installing sheave hold-down hardware and the friction-shoe assembly. When assembled correctly, the sharp edges of the friction shoes are poised for contact with the flywheel-hub inside diameter (Fig. 5-17F).
7. Pull the centering pin out about 1/8 in. for positive engagement with the crankshaft center hole.
8. Install the starter on the blower housing, rotating the flywheel with the starter rope as the hold-down screws are torqued. This procedure helps to center the clutch in the flywheel hub.
9. Start the engine to verify starter operation.

The Fairbanks-Morse utility starter is a smaller and simpler version of the heavy-duty models just discussed (Fig. 5-18). A one-piece sheave is used with the rope anchored by a knot, rather than a compression fitting. The utility starter uses the same clutch components as its larger counterparts and, like them, can be assembled for right- or left-hand engine rotation.

Vertical pull

Vertical-pull starters are an area where DIY mechanics shine. These starters are obsolete, complicated, and troublesome. Commercial shops usually won't fool with them and when they do, the labor charges can be horrendous. But a do-it-yourselfer can, with a bit of patience, repair these starters and, in the process, salvage engines that would otherwise be scrapped.

Like other spring-powered devices, these starters must be disarmed before disassembly. Otherwise, the starter will disarm itself with unpredictable results. Disarming involves three distinct steps: releasing mainspring pretension (usually by slipping a foot or so of rope out of the sheave flange and allowing the sheave to unwind), disengaging the mainspring anchor (usually held by a threaded fastener), and when the spring is to be replaced, uncoiling the spring from its housing.

Warning: Safety glasses are mandatory for disassembly.

Vertical-pull starters tend to be mechanically complex and—because of a heavy reliance upon plastic, light-gauge steel, and spring wire—are unforgiving. Parts easily bend or break. Lay components out on the bench in proper orientation and in sequence of disassembly. If there is any likelihood of confusion, make sketches to guide assembly. Also, note that the step-by-step instructions in this book must aim at thoroughness and describe all operations, but it will rarely be necessary to follow every step and completely dismantle a starter.

Briggs & Stratton

Briggs & Stratton has used one basic vertical-pull starter with minor variations in the link and sheave mechanisms. It is probably the most reliable of these starters, and the easiest to repair.

Disassembly

1. Remove starter assembly from the engine.
2. Release mainspring pretension by lifting the rope out of the sheave flange and, using the rope for purchase, winding the sheave counterclockwise two or three revolutions (Fig. 5-19).
3. Carefully pry the plastic cover off with a screwdriver. Do not pull on the rope with the cover off and spring anchor attached; doing so can permit the outer end of the spring to escape the housing.
4. Remove the spring anchor bolt and spring anchor (Fig. 5-20). If the mainspring is to be replaced, carefully extract it from the housing,

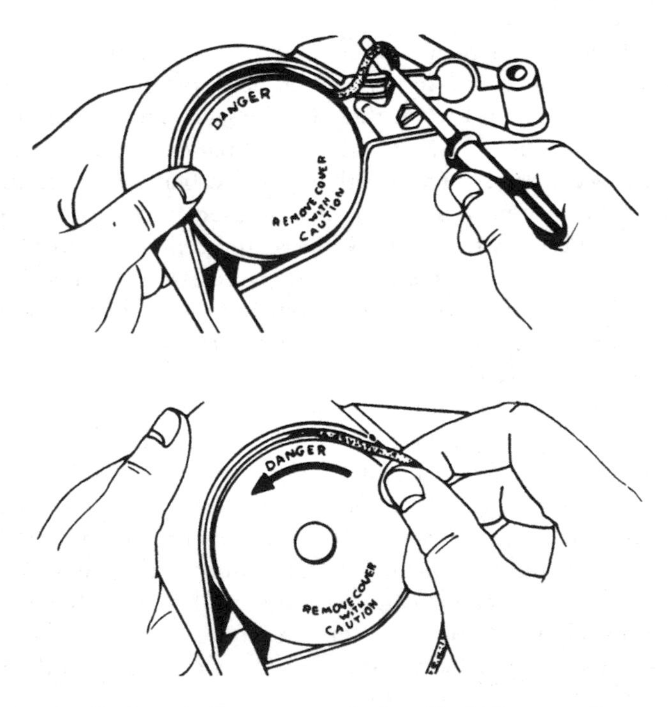

FIGURE 5-19. *Briggs & Stratton vertical-pull starters are disarmed by slipping the rope out of the sheave groove and using the rope to turn the sheave two or three revolutions counterclockwise until the mainspring relaxes.*

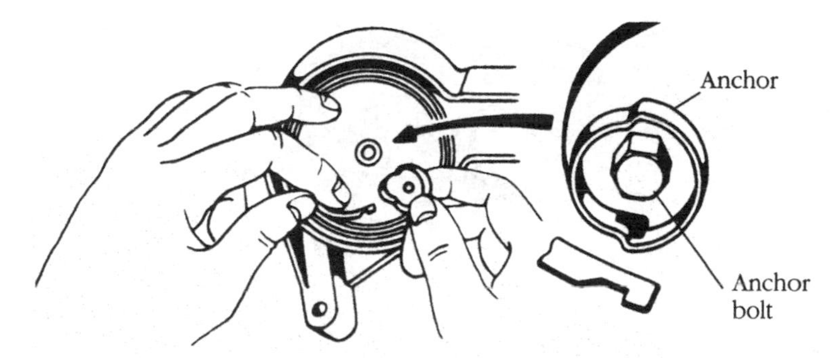

FIGURE 5-20. *Mainspring anchor bolt must be torqued 75 lb/in. and can be further secured with thread adhesive.*

working from the center coil outward. Note the spring lay for future reference.

5. Separate the sheave and the pin (Fig. 5-21). Observe the link orientation.
6. The rope can be detached from the sheave with the aid of long-nosed pliers. Figure 5-22 shows this operation and link retainer variations.
7. The rope can be disengaged from the handle by prying the handle center section free and cutting the knot (Fig. 5-23).
8. Clean all parts (except rope) in petroleum-based solvent to remove all traces of lubricant.
9. Verify the gear response to link movement as shown in Figure 5-24. The gear should move easily between its travel limits. Replace the link as necessary.

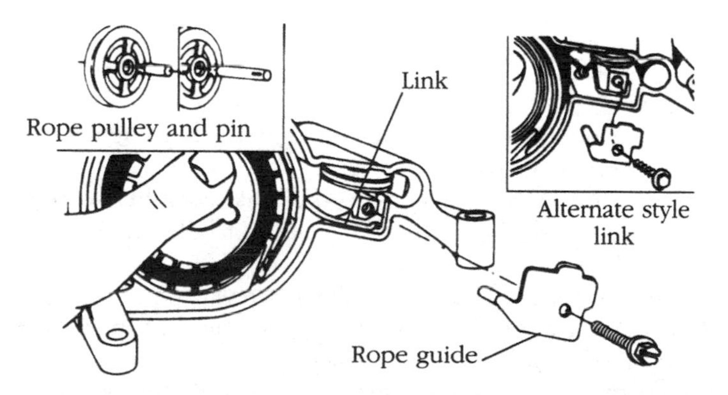

FIGURE 5-21. *Make note of the friction-link orientation for assembly.*

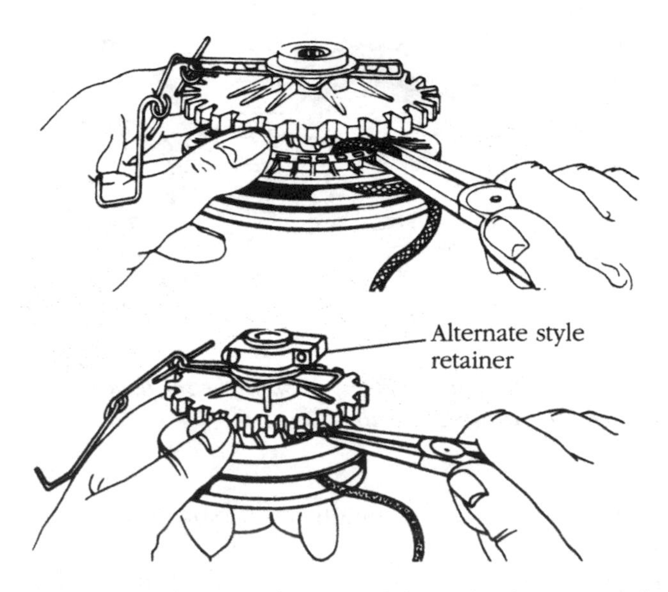

Figure 5-22. *The rope can be disengaged from the sheave with long-nosed pliers.*

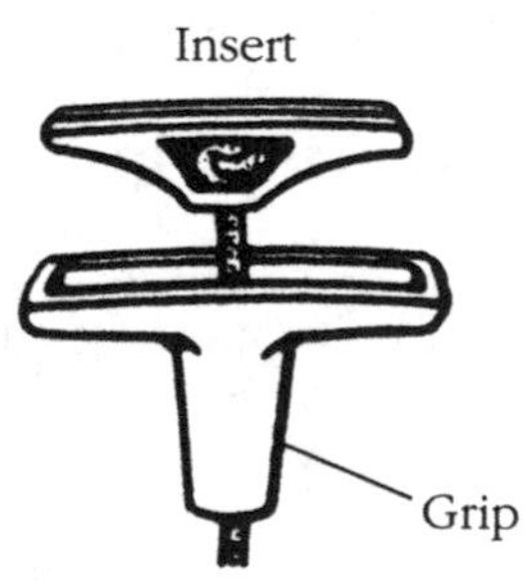

Figure 5-23. *The Briggs handle employs an insert that must be extracted to renew the rope.*

Assembly

1. Install the outer end of the mainspring in the housing retainer slot and wind counterclockwise (Fig. 5-25).
2. Mount the sheave, sheave pin, and link assembly in the housing. Index the end of the link in the groove or hole provided (Figs. 5-26 and 5-27).
3. Install the rope guide and hold-down screw.
4. Rotate the sheave counterclockwise, winding the rope over the sheave (Fig. 5-28).
5. Engage the inner end of the mainspring on the spring anchor. Mount the anchor and torque the hold-down caps crew to 75–90 lb/in.

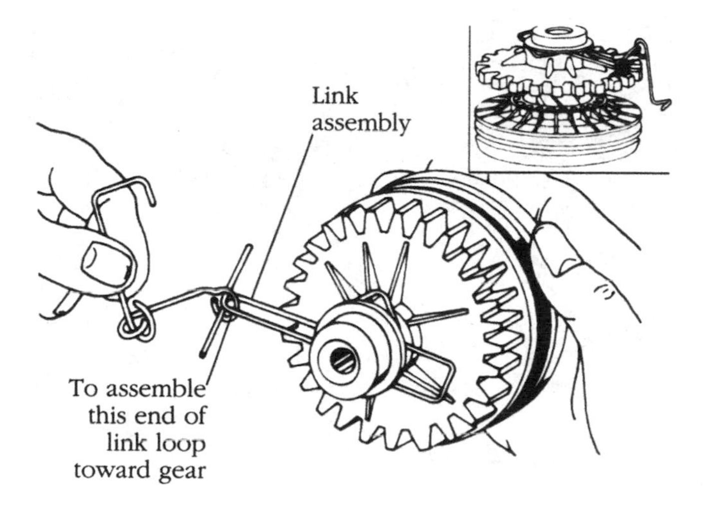

FIGURE 5-24. *The pinion gear should move through its full range of travel in response to link movement. The inset on the upper right of the illustration shows link orientation.*

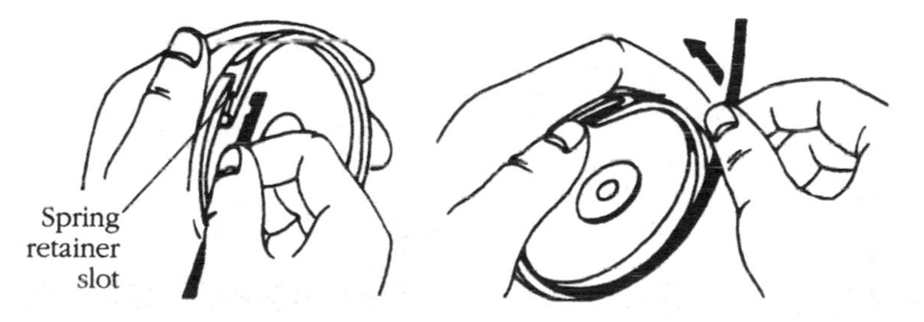

FIGURE 5-25. *The mainspring winds counterclockwise from the outer coil.*

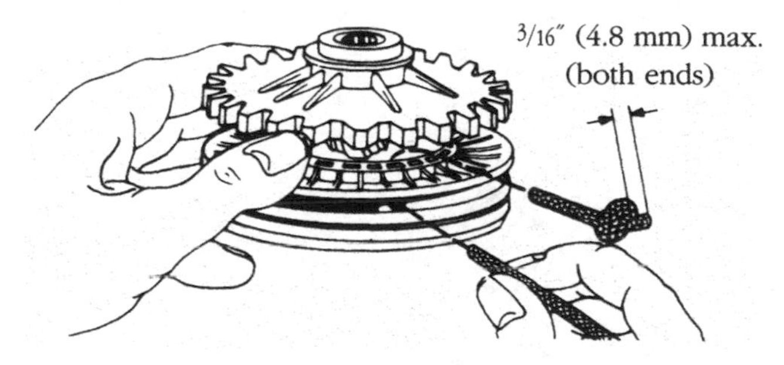

FIGURE 5-26. *A short length of piano wire aids rope insertion into the sheave.*

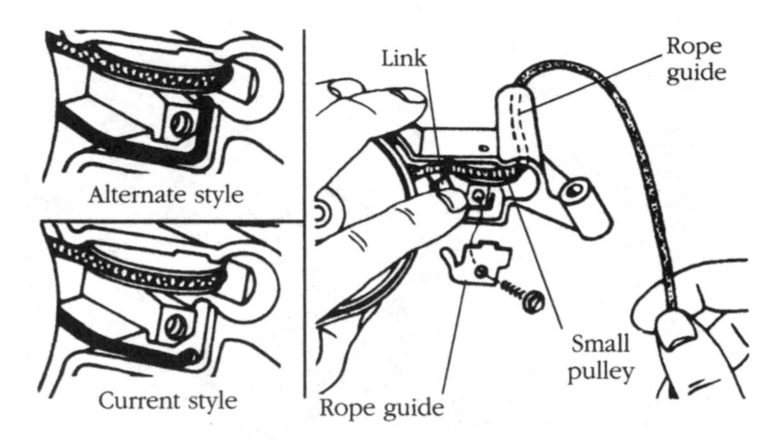

FIGURE 5-27. *Friction link hold-down detail.*

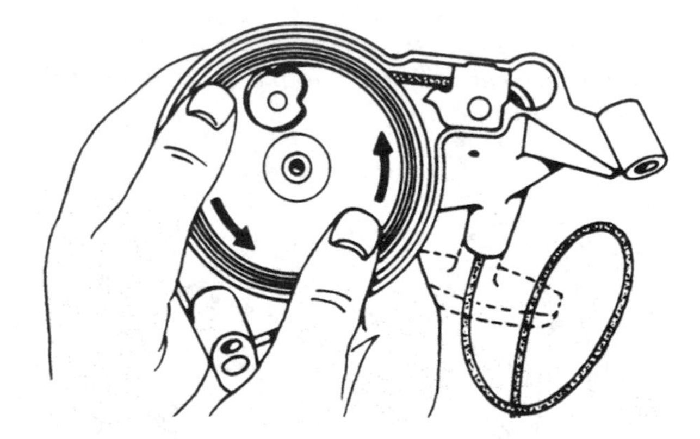

FIGURE 5-28. *The rope winds counterclockwise on the sheave, then the spring anchor and anchor bolt are installed.*

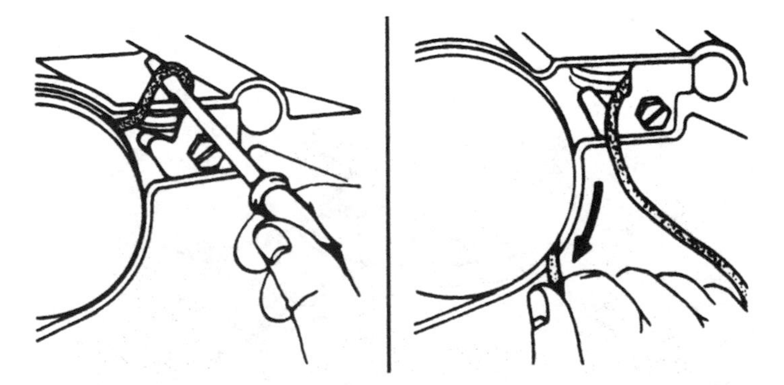

FIGURE 5-29. *Pretension requires two or three sheave revolutions using the rope for leverage.*

6. Snap the plastic cover into place over the spring cavity.
7. Disengage 12 in. or so of rope from the sheave and, using the rope for purchase, turn the sheave two or three revolutions clockwise to generate pretension (Fig. 5-29).
8. Mount the starter on the engine and test.

Tecumseh

Tecumseh has used several vertical-pull starters, ranging from quickie adaptations of side-pull designs to the more recent vertical-engagement type.

The gear-driven starter shown in Figure 5-30 is an interesting transition from side to vertical-pull. No special service instructions seem appropriate, except to provide plenty of grease in the gear housing and some light lubrication on the mainspring. Assemble the brake spring without lubricant.

The current horizontal engagement starter (Fig. 5-31) is reminiscent of the Briggs & Stratton design, with a rope clip, cup-type spring anchor ("hub" in the drawing), and threaded sheave extension upon which the pinion rides.

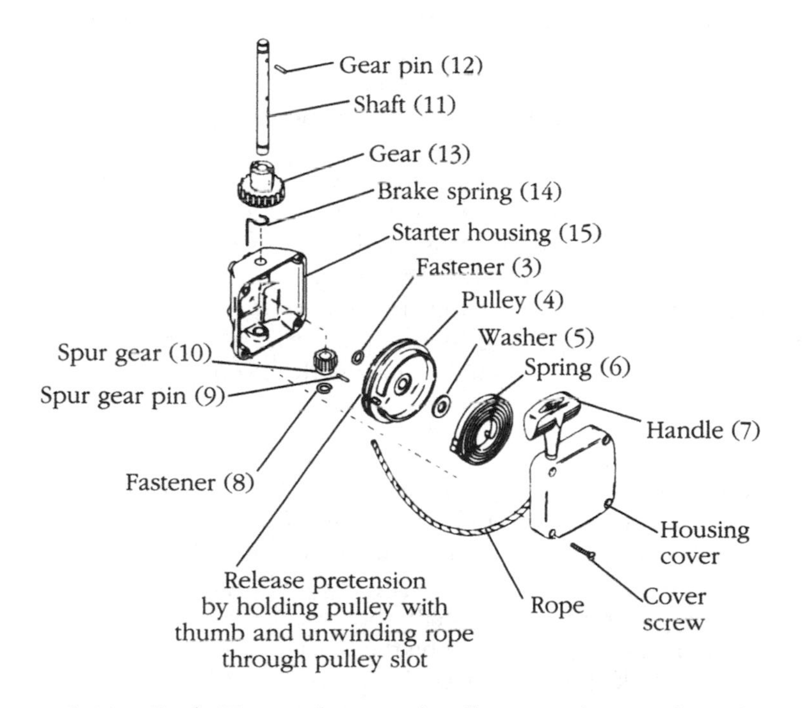

FIGURE 5-30. *Early Tecumseh vertical-pull starter, driving through a gear train. While heavy (and, no doubt, expensive to manufacture), this starter was quite reliable.*

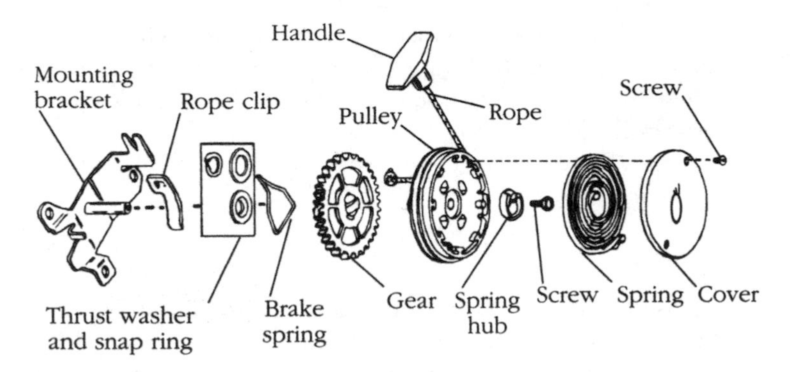

FIGURE 5-31. *Tecumseh's most widely used vertical-pull starter employed a spiral gear to translate the pinion horizontally into contact with the flywheel.*

Disassembly

1. Remove the unit from the engine.
2. Detach the handle and allow the rope to retract past the rope clip. This operation relieves mainspring preload tension.
3. Remove the two cover screws and carefully pry the cover free.
4. Remove the central hold-down screw and spring hub.
5. Protecting your eyes with safety glasses, extract the mainspring from the housing. Work the spring free a coil at a time from the center out. If the spring will be reused, it can remain undisturbed.
6. Lift off the gear and pulley assembly. Disengage the gear and, if necessary, remove the rope from the pulley.
7. Clean all parts except the rope in solvent.
8. Inspect the brake spring (the Achilles' heel of vertical-pull starters). The spring must be in solid contact with the groove in the gear.

Assembly

1. Secure the rope to the handle, using No. 4 1/2 or 5 nylon rope, 61 in. long for standard starter configurations. Cauterize the rope ends and form by wiping with a cloth while the rope is still hot.
2. Assemble the gear on the pulley, using no lubricant.
3. Lightly grease the center shaft and install the gear and pulley. The brake spring loop is secured by the bracket tab. The rope clip indexes with the hole in the bracket (Fig. 5-32).

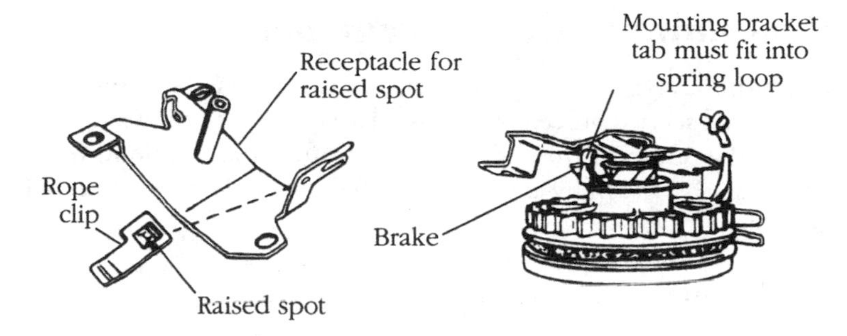

FIGURE 5-32. *The rope clip and spring loop index to the bracket.*

4. Install the hub and torque the center screw to 44–55 lb/in.
5. Install the spring. New springs are packed in a retainer clip to make installation easier.
6. Install the cover and cover screws.
7. Wind the rope on the pulley by slipping it past the rope clip. When fully wound, turn the pulley two additional revolutions for preload.
8. Mount the starter on the engine, adjusting the bracket for minimum 1/16-in. tooth clearance (Fig. 5-33). Less clearance could prevent disengagement, destroying the starter.

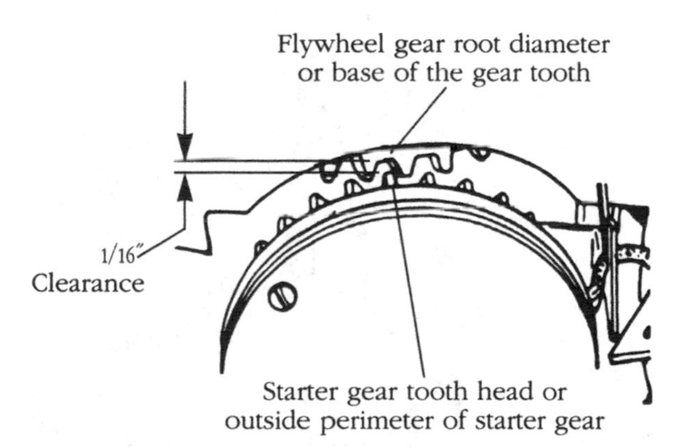

FIGURE 5-33. *Generous gear lash, minimum 1/16 in., is required to assure pinion disengagement when the engine starts.*

Vertical pull, vertical engagement

The vertical-pull, vertical-engagement starter is a serious piece of work that demands special service procedures. It is relatively easy to disassemble while still armed. The results of this error can be painful. Another point to note is that rope-to-sheave assembly as done in the field varies from the original factory assembly.

Rope replacement

Figure 5-34 is a composite drawing of several vertical-pull starters. Many do not contain the asterisked parts, and early models do not have the V-shaped groove on the upper edge of the bracket that simplifies rope replacement.

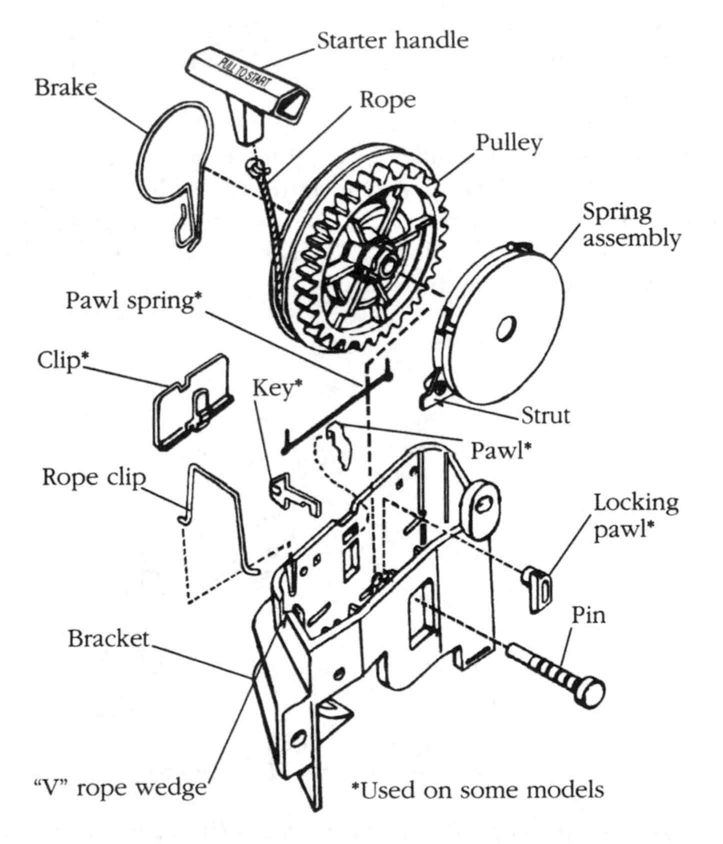

FIGURE 5-34. *Tecumesh's vertical-pull, vertical-engagement starter is the most sophisticated unit used on small engines. The mainspring and its retainer are integral and are not separated for service.*

When this groove is present, the rope (No. 4 1/2, 65-in. standard length, longer with a remote rope handle) can be renewed by turning the sheave until the staple, which holds the rope to the sheave, is visible at the groove (Fig. 5-35). Pry out the staple and wind the sheave tight. Release the sheave a half turn to index the hole in the sheave with the V-groove. Insert one end of the replacement rope through the hole, out through the bracket. Cauterize and knot the short end, and pull the rope through, burying the knot in the sheave cavity. Install the rope handle, replacing the original staple with a knot, and release the sheave. The rope should wind itself into place.

Disassembly

1. Remove the starter from the engine.
2. Pull out the rope far enough to secure it in the V-wedge on the bracket end. This part, distinguished from the V-groove mentioned above, is called out in Figure 5-34.
3. The rope handle can be removed by prying out the staple with a small screwdriver.
4. Press out the flat-headed pin that supports the sheave and spring the capsule in the bracket. This operation can be done in a vise with a large deep well socket wrench as backup.
5. Turn the spring capsule to align with the brake spring legs. Insert a nail or short (3/4-in. long maximum) pin through the hole in the strut and into the gear teeth (Fig. 5-36).

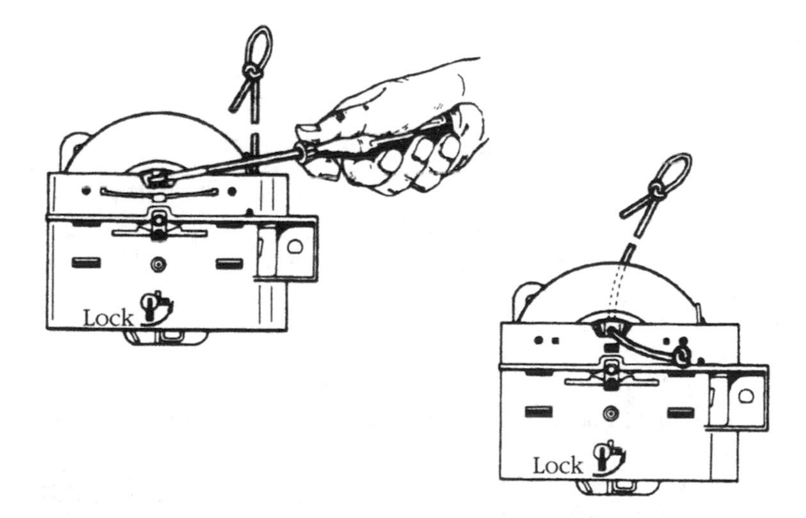

FIGURE 5-35. *V-groove in the bracket gives access to the rope anchor on some models.*

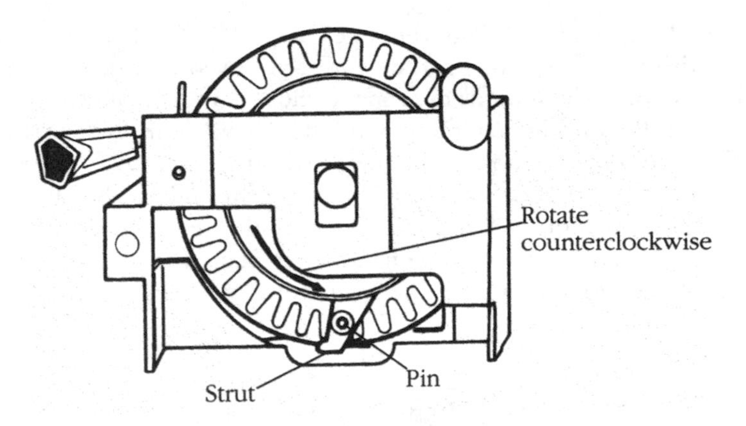

FIGURE 5-36. *A pin locks the spring capsule and gear to prevent sudden release of the mainspring tension.*

6. Lift the sheave assembly and pry the capsule out of the bracket. Warning: Do not separate the sheave assembly and capsule until the mainspring is completely disarmed.
7. Hold the spring capsule firmly against the outer edge of the sheave with thumb pressure and extract the locking pin inserted in Step 5.
8. Relax pressure on the spring capsule, allowing the capsule to rotate, thus dissipating residual mainspring tension.
9. Separate the capsule from the sheave and, if rope replacement is in order, remove the hold-down staple from the sheave.
10. Clean and inspect all parts.
    Note: No lubricant is used on any part of this starter.

Assembly

1. Cauterize and form the ends of the replacement rope (see specs above) by wiping down the rope with a rag while still hot.
2. Insert one end of the rope into the sheave, 180° away from the original (staple) mount (Fig. 5-37A).
3. Tie a knot and pull the rope into the knot cavity.
4. Install the handle (Fig. 5-37B).
5. Wind the rope clockwise (as viewed from the gear) over the sheave.
6. Install the brake spring, spreading the spring ends no more than necessary.
7. Position the spring capsule on the sheave, making certain the mainspring end engages the gear hub (Fig. 5-38A).
8. Wind four revolutions, align the brake spring ends with the strut (Fig. 5-38B), and lock with the pin used during disassembly.

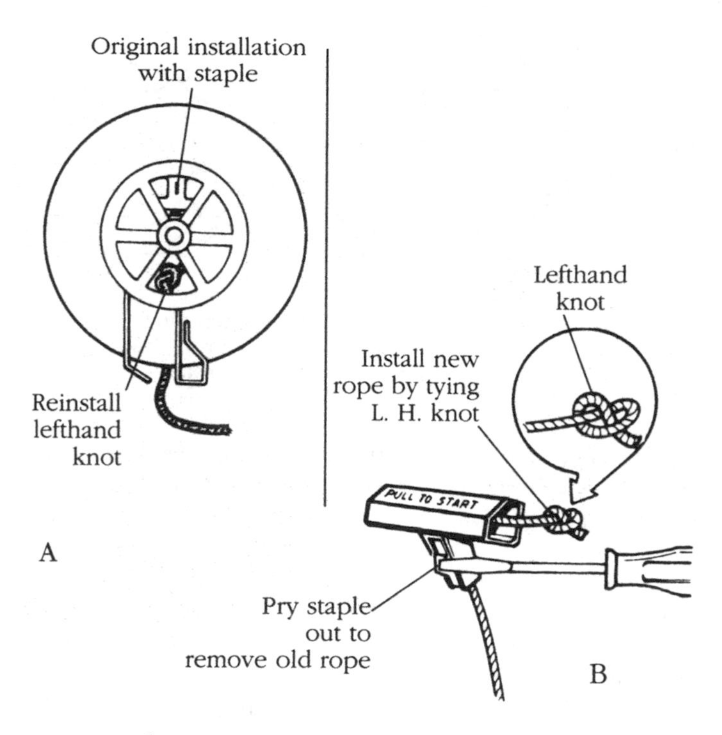

FIGURE 5-37. *Replacement ropes anchor with a knot, rather than staple, and mount 180° from the original position on the sheave.*

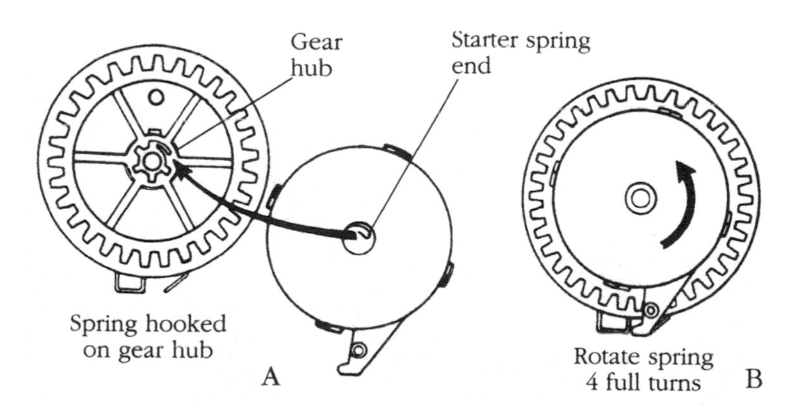

FIGURE 5-38. *The spring capsule engages with the gear hub (A), is rotated four revolutions for pretension, and pinned (B).*

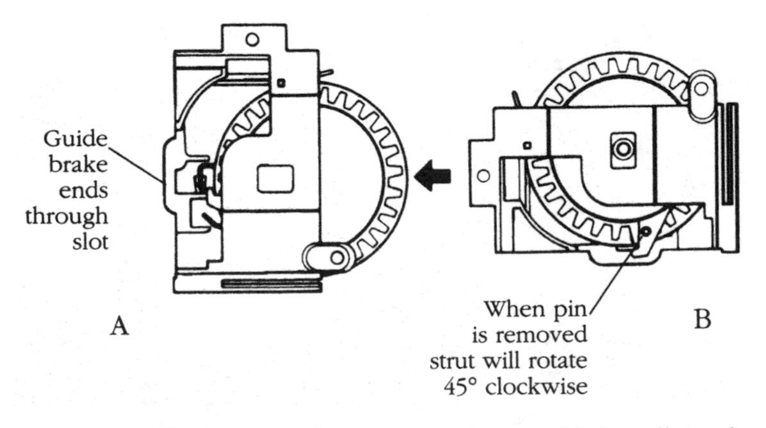

FIGURE 5-39. *The sheave and spring capsule assembly installs in the bracket with the brake spring ends in slots (A). Releasing pin arms starter (B), which can now be mounted on the engine.*

9. Install pawls, springs, and other hardware that might be present.
10. Insert the sheave and spring assembly into the bracket, with the brake spring legs in the bracket slots (Fig. 5-39).
11. Feed the rope under the guide and snub it in the V-notch.
12. Remove the locking pin, allowing the strut to rotate clockwise until retained by bracket.
13. Press or drive the center pin home.
14. Mount the starter on the engine and test.

6

Electrical system

At its most developed, the electrical system consists of a charging circuit, a storage battery, and a starting circuit. A flywheel alternator provides electrical energy that is collected in the battery for eventual consumption by the starter motor.

Not all small engine electrical systems include both circuits. Some dispense with the starting circuit and others employ a starting circuit without provision for onboard power generation.

Starting circuits

Starting circuits fall into two major groups: dc (direct current) systems that receive power from a 6 or 12 V battery and ac (alternating current) systems that feed from an external 120 Vac line. I will not discuss ac systems because the hazards implicit in line-current devices cannot be adequately addressed in a book of this type. My discussion is limited to dc systems that employ conventional (lead-acid) or NiCad batteries.

Lead acid

As shown in Figures 6-1 and 6-2, a conventional starting circuit includes four major components—battery, ignition switch, solenoid, and motor—wired into two circuit loops:

- Control loop—14-gauge primary wire from the positive battery terminal, through the ignition switch, to the solenoid windings.
- Power loop—cable from the positive battery terminal, through the solenoid and to the starter motor.

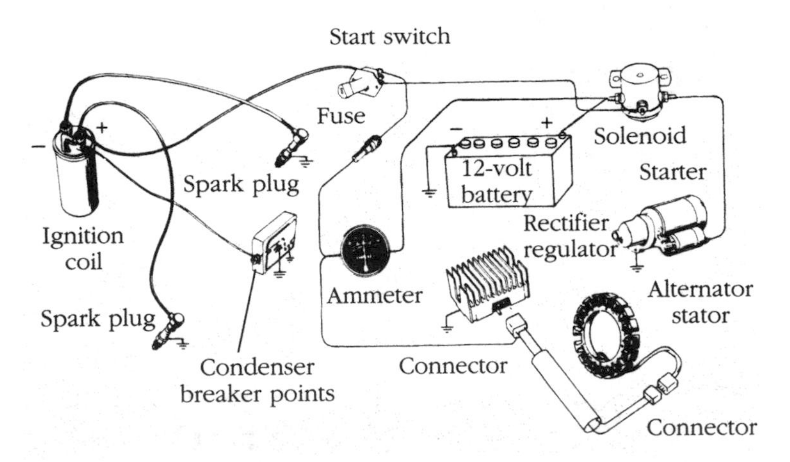

FIGURE 6-1. *Electrical system supplied with Onan engines and tied into battery-and-coil ignition. Note the heavy-duty stator and combined rectifier/regulator.*

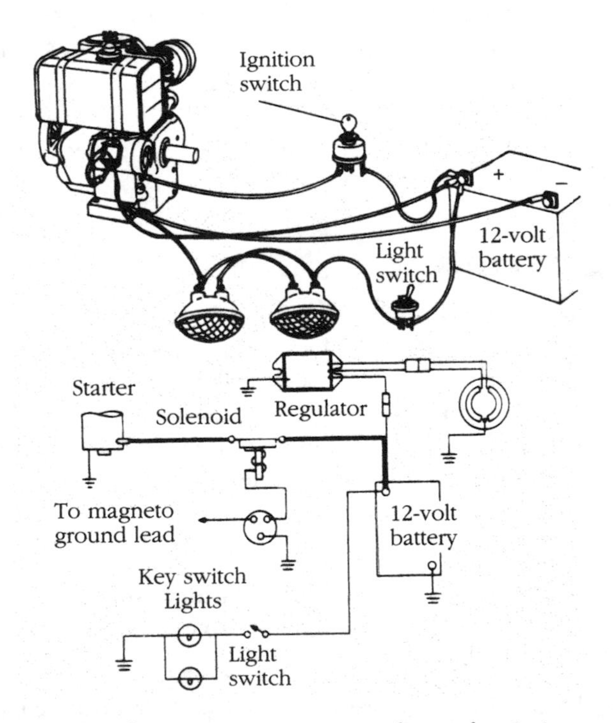

FIGURE 6-2. *Briggs & Stratton 10A system shows the tie-in to a charging system on the positive battery post. The ignition switch grounds magneto and does not interchange with automotive-type switches.*

As shown in the drawings, the negative battery terminal connects to the engine block to provide a ground return for both loops. When energized, current flows from the positive side of the solenoid, starter motor, and other circuit components to ground.

The solenoid—more properly called a relay—is a normally open (NO) electromagnetic switch. When energized by the starter switch, the solenoid closes with an audible click to complete the power circuit. Most solenoids are internally grounded, which means that mounting faces must be clean and hold-down bolts secure. Interlocks are sometimes included in the control circuit to prevent starting under unsafe conditions (e.g., if the machine is in gear). The charging system delivers current to the positive post of the battery.

Figure 6-3 outlines diagnostic procedures which use the solenoid as the point of entry. Shunting the solenoid with a jumper cable removes the solenoid and the control loop from the circuit to give an immediate indication of starter motor function.

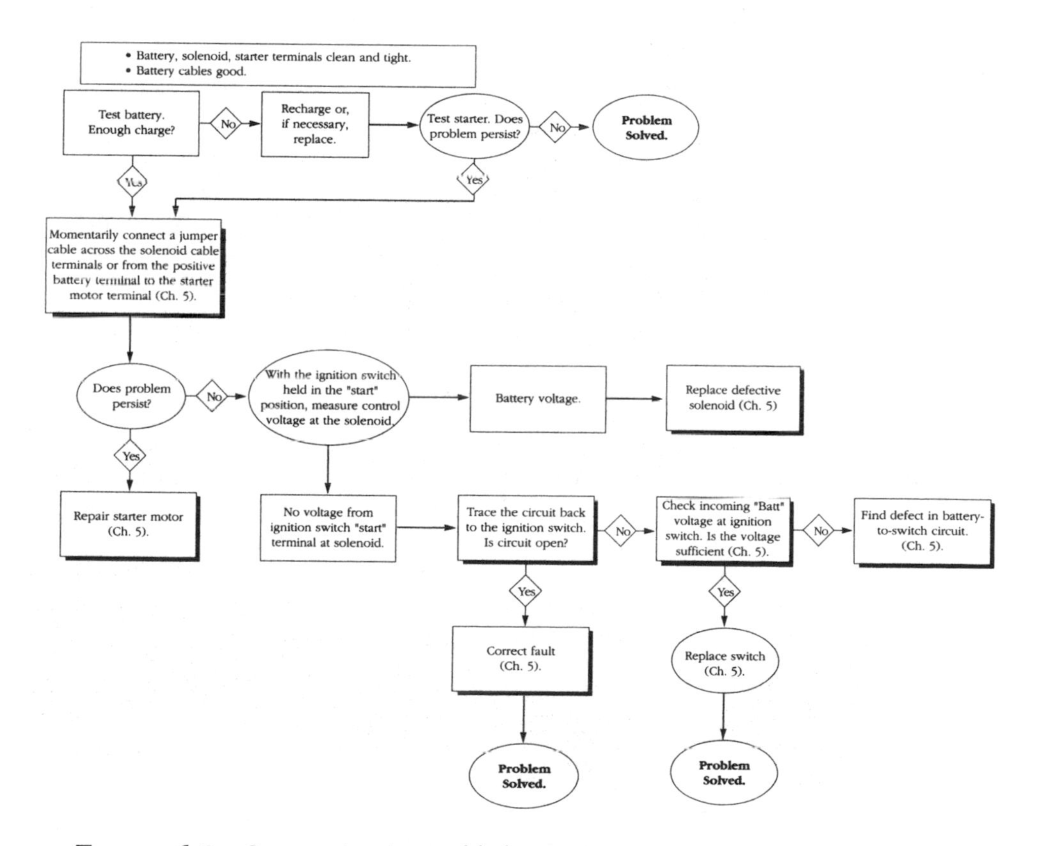

FIGURE 6-3. *Starter circuit troubleshooting.*

The majority of starting-circuit malfunctions are the fault of the battery. Loose, dirty, or corroded connections account for most of the others.

1. Remove the cable connections at the battery and scrape the battery terminals to bright metal. Repeat the process for solenoid and starter-motor connections.
2. Clean the battery connection at the engine and verify that all circuit components are bolted down securely to the engine or equipment frame.
3. Verify that electrolyte covers the battery plates. Add distilled water as necessary, but do not overfill.
4. Clean the battery top with a mixture of baking soda, detergent, and water. Rinse with fresh water and wipe dry.
5. With the battery cables disconnected, charge the battery.

Warning: Lead-acid batteries give off hydrogen gas during charging. To minimize sparking and possible explosion, connect the charger cables (red to positive, black to negative) before switching ON the charger; switch OFF the charger before disconnecting.

6. Test each cell with a hydrometer. Replace the battery if the charger cannot raise the average cell reading to at least 1.260 or if individual cell readings vary more than 0.050.
7. Connect a voltmeter across the battery terminals. With ignition output grounded, crank the engine for a few seconds. Cranking voltage should remain above 9.5 V (12 V systems) or 4.5 V (6 V). Lower readings mean a defective battery or starter circuit.

Caution: Small-engine starters have limited duty cycles. Allow several minutes for the starter to cool between ten-second cranking periods.

NiCad

As far as I am aware, only two manufacturers—Briggs & Stratton and Tecumseh—supply NiCad-powered systems. As show in Figure 6-4, the circuit includes a nickel-cadmium battery pack, a switch, and a 12 Vdc starter motor, all specific to these systems and not interchangeable with any other. A 110 Vac charger replenishes the battery pack before use.

Troubleshooting diagnostic procedures are straightforward:

1. Verify the ability of the battery pack to hold a charge. If necessary, test the 110 Vac charger.
2. If a known-good battery pack does not function on the engine, check the control switch with an ohmmeter.
3. Test the starter motor.

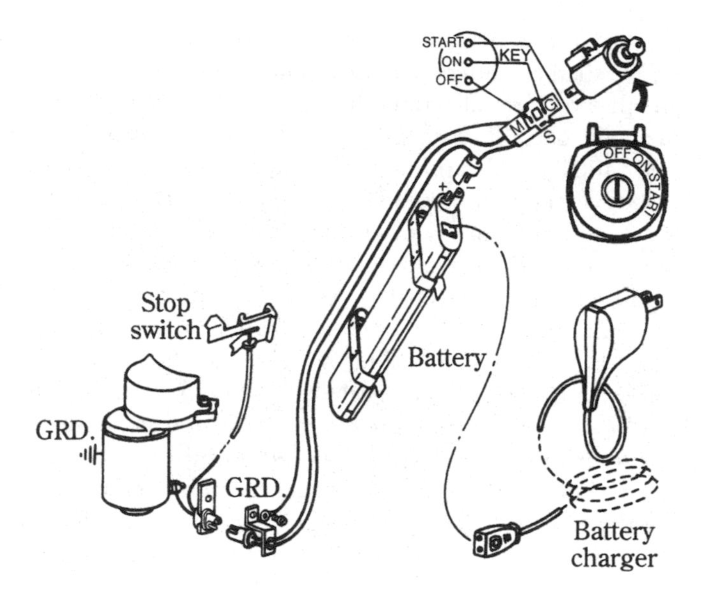

FIGURE 6-4. *Briggs & Stratton Nicad system integrates engine controls with the wiring harness. Tecumseh's practice is similar.*

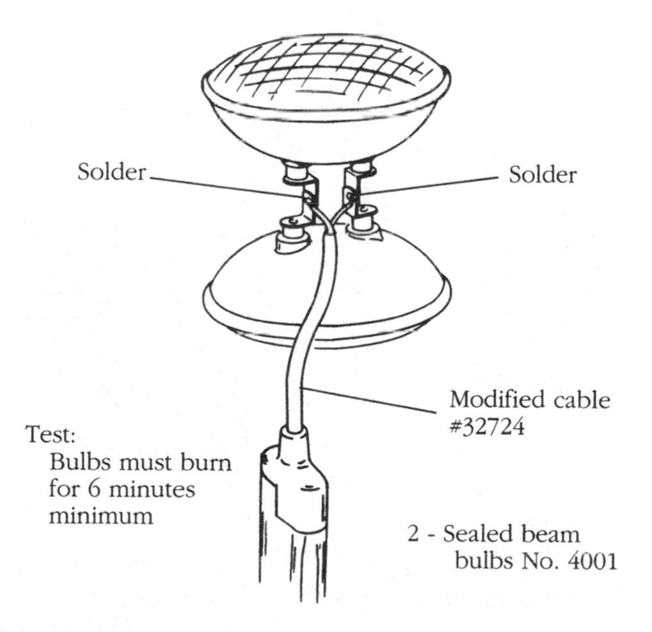

FIGURE 6-5. *Nicad test load is fabricated from two sealed-beam headlamps and a battery-to-starter cable.*

Refusal of the battery pack to hold a charge is the most common fault. After 16 hours on the charger, battery potential should range between 15.5 V and 18 V. Assuming that voltages fall within these limits, the next step is to test capacity through a controlled discharge. If a carbon-pile tester is not available, connect two No. 4001 headlamp bulbs in parallel, as shown in Figure 6-5. A freshly charged battery pack should illuminate the lamps brightly for five minutes (Briggs & Stratton) or six minutes (Tecumseh), figures that represent enough energy to start the average engine about 30 times.

Warning: Dispose of NiCad battery packs in a manner approved by local authorities. Cadmium, visible as a white powder on leaking cells, is a persistent poison. Do not incinerate and/or weld near the battery pack.

Battery life will be extended if charging is limited to 12 or 16 hours immediately before use and once every two months in dead storage.

NiCad charger output varies with battery condition, but after two or three hours it should be about 80 mA. Tecumseh lists a test meter (PN 670235) for the PN 32659 charger; Briggs suggests that the technician construct a tester, as described in Figure 6-6.

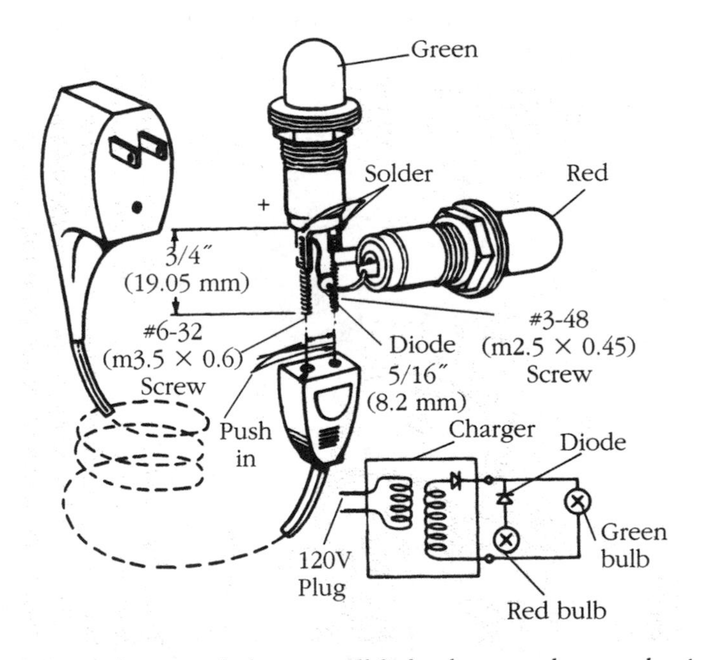

FIGURE 6-6. *A functional charger will light the green lamp only. A charger with an open diode will light the red bulb; one with a shorted diode will light both bulbs. Parts required: one 1N4005 diode, two Dialco lamp sockets (PN 0931-102 red, PN 0932-102 green), two No. 53 bulbs, and hold-down screws.*

Starter motors

Industrial motors, such as the Prestolite unit shown in Figure 6-7, are rebuildable and can be serviced by most automotive electrical shops. The Briggs & Stratton motor, shown in Figure 6-8, is also rebuildable (thanks to adequate parts support) and can, in the context of small engines, be considered a heavy-duty motor. American Bosch, used by Kohler and shown in Figure 6-9, is, from a reparability point of view, on a par with the Briggs. European Bosch, Bendix, Nippon Denso, and Mitsubishi starter motors are of similar quality. Light-duty units, such as the NiCad starter shown in Figure 6-10, do not justify serious repair efforts.

Troubleshooting. Figure 6-11 describes motor failure modes and likely causes. References to field failures apply only to those motors that use wound field coils; PM (permanent magnet) fields are, of course, immune to electrical malfunction.

Repairs. With the exception of replacing the inertial clutch, repair procedures discussed here apply to heavy-duty motors. Upon disassembly, clean the interior of the starter with an aerosol product intended for this purpose. Do not use a petroleum-based solvent. Note the placement of thrust and insulating washers.

- Inertial clutch—Shown clearly in Figure 6-10 and tangentially in other drawings, the Bendix is serviced as a complete assembly. It secures to the motor shaft with a nut, spring, clip, or roll pin. Support the free end of the motor shaft when driving the pin in or out. The helix and gear install dry, without lubrication; the pinion ratchet can be lightly oiled.
- End cap—Scribe the end cap and motor frame as an assembly aid. Installation can be tricky when the brush assembly is part of the cap. In some cases, you can retract the brushes with a small screwdriver. Radially deployed brushes can be retained with a fabricated bracket (Fig. 6-12)
- Bushings—Do not disturb the bushings unless replacements are at hand. Drive out the pinion-end bushing with a punch sized to the bushing outside diameter. Commutator bushings must be lifted out of their blind end-cap bosses, which can be done by filling the cavity with grease and using a punch, sized to match the shaft outside diameter as a piston. Hammer the punch into the grease.

  Sintered bronze bushings—recognizable by their dull, sponge-like appearance, these bushings should be submerged in motor oil for a few minutes before installation. Brass bushings require a light temperature-resistant grease, such as Lubriplate.

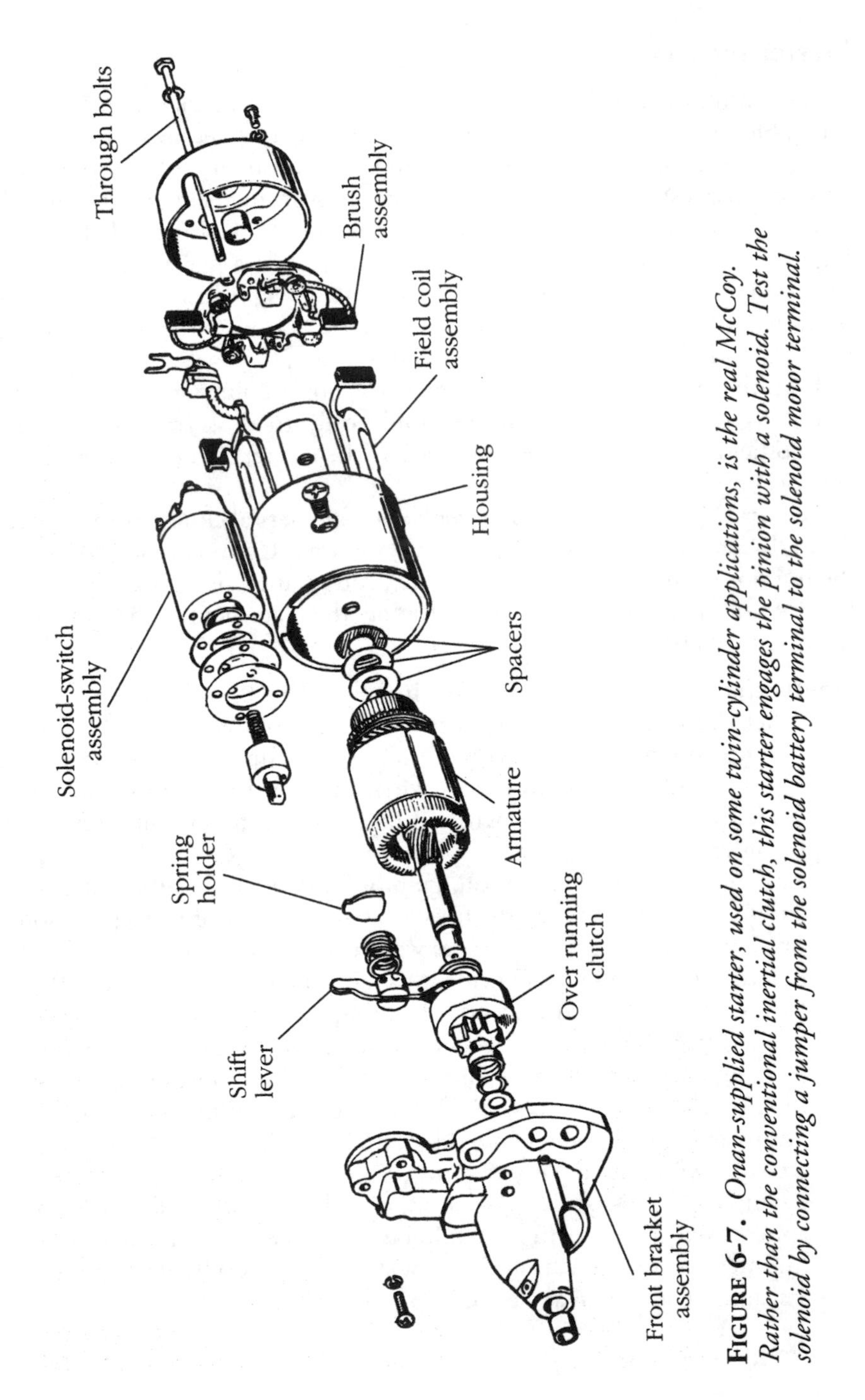

FIGURE 6-7. *Onan-supplied starter, used on some twin-cylinder applications, is the real McCoy. Rather than the conventional inertial clutch, this starter engages the pinion with a solenoid. Test the solenoid by connecting a jumper from the solenoid battery terminal to the solenoid motor terminal.*

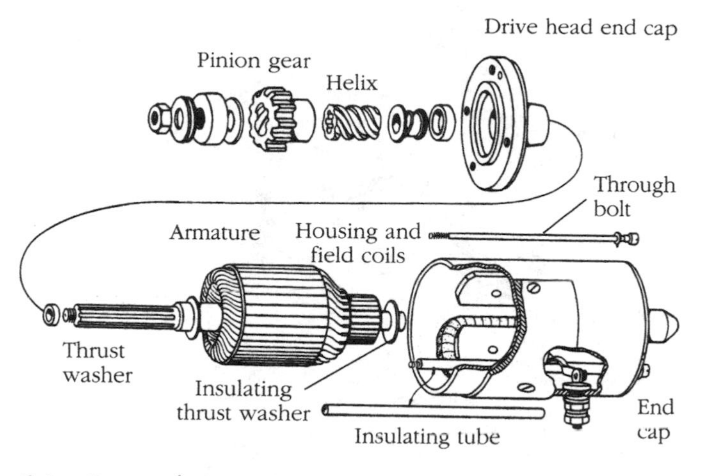

FIGURE 6-8. *Briggs & Stratton 12 Vdc starter motor employs electromagnetic (EM) fields, a thrust washer on the drive end, and an insulating thrust washer at the commutator.*

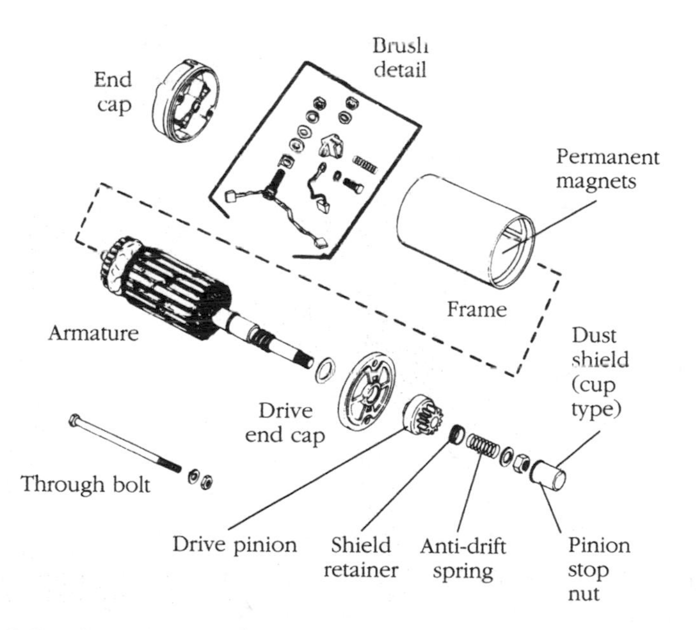

FIGURE 6-9. *American Bosch 12 Vdc starter features permanent magnet (PM) fields and a radial commutator with brushes parallel to the motor shaft. Used on Kohler and other serious engines.*

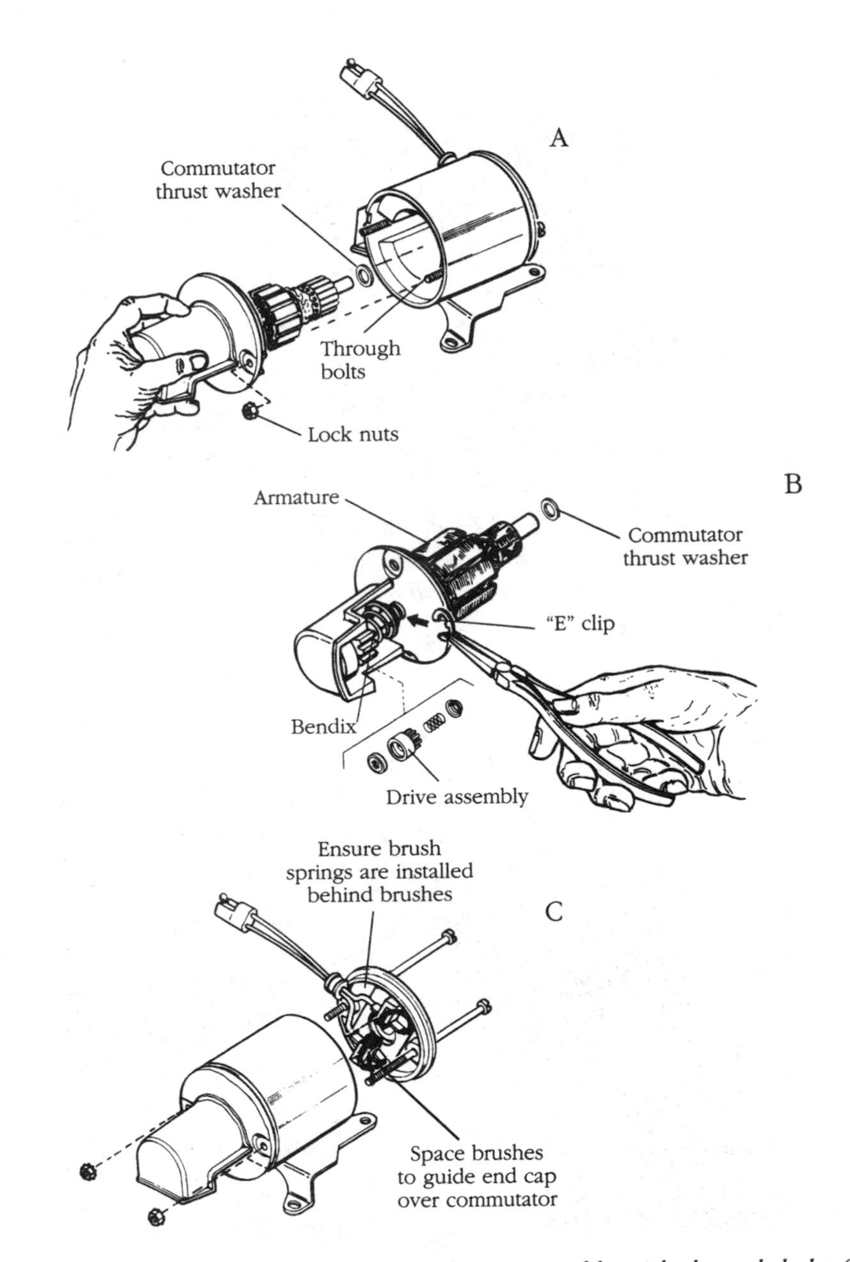

FIGURE 6-10. *Tecumseh NiCad end caps assemble with through bolts (A). Bendix inertial clutch secures with an E-clip (B). Note the thrust washer (C). Brushes, replaceable only as part of the cap assembly, must be shoehorned over the commutator. This particular starter should draw 20A while turning the engine 415 rpm with crankcase oil at room temperature.*

It is assumed that full battery voltage reaches the starter and that the flywheel offers only normal resistance to turning.

Starter does not function:

- Brushes stuck in holders
- Dirty, oily brushes/commutator
- Open, internally shorted, or grounded field coil
- Open, internally shorted, or grounded armature

Starter cranks slowly
(minimal acceptable cranking speed = 350 rpm):

- Worn brushes or weak brush springs
- Dirty, oily, or worn commutator
- Worn shaft bushings
- Defective armature

Starter stalls under compression:

- Overly advanced ignition timing
- Defective armature
- Defective field coil

Starter works intermittently:

- Sticking brushes
- Loose connections in external circuit
- Dirty, oily commutator

Starter spins freely without turning flywheel:

- Pinion gear sticking on shaft
- Broken pinion and/or flywheel teeth

FIGURE 6-11. *Starter motor faults and probable causes.*

- Brushes—Most starter problems originate with brushes that wear short and bind against the sides of their holders. As a rule of thumb, replace the brushes when worn to half their original length. Older starters used screw-type brush terminals; newer starters employ silver solder or integrate brushes and brush holders with the cap. Note the lay of the brushes: Rubbing surfaces must conform to the convexity of the commutator.
- Commutator—Heavy-duty starters can be "skimmed" on a lathe to restore commutator concentricity and surface finish. A light-duty commutator might benefit from polishing with 000-grade sandpaper. Do not use emery cloth.

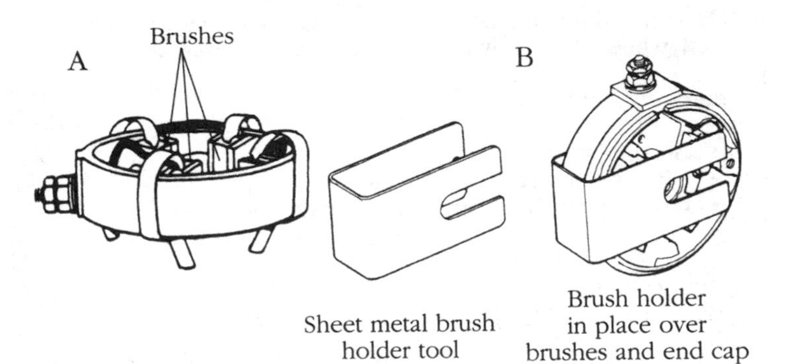

FIGURE 6-12. *Fabricated brush holders for four-pole (brush) assemblies (A) and two-pole (B).*

- Armature—Check continuity with an ohmmeter or 12 V trouble light. Two conditions must be met:
  (1) Paired commutator bars are connected in series and should provide a continuous circuit path from a brush on one side of the commutator to its twin on the other side.
  (2) No pair of bars has continuity with other pairs or with the motor shaft (Fig. 6-13).

  Automotive electrical shops can test the armature for internal shorts with a growler.

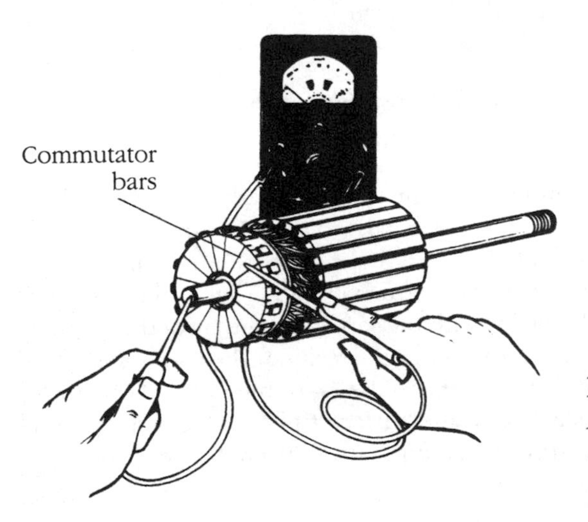

FIGURE 6-13. *Testing an American Bosch starter commutator for shorts to the motor shaft.*

- Fields—inspect PM fields for mechanical damage (from contact with the armature) and for failure of the adhesive backing. A specialist should test electromagnetic fields, although at this point you have reached the end of practical reparability for even the best starter motor.

Charging circuits

In its most vestigial form, a charging circuit consists of a coil, flywheel magnet, and a load, such as a headlamp. Coil output alternates, or changes direction, each time a flywheel magnet excites it (the discussion of magneto theory in Chap. 3 explains why). Voltage is speed-sensitive: at idle the lamp barely glows; at wide-open throttle the filament verges on self-destruct.

Adding a battery means that stator output must be rectified, or converted from ac to dc. This is almost always done by means of one or two silicon diodes, which act as check valves to pass current flowing in one direction and block it in the other. Single-diode rectifiers pass that half of stator output that flows in the favored direction (Fig. 6-14A). Full-wave rectifiers use two diodes, wired in a bridge circuit, to impose unidirectionality upon all of the output, so that none of it goes to waste (Fig. 6-14B).

The battery receives a charge so long as its terminal voltage is lower than rectifier output voltage. The battery also acts as a ballast resistor, limiting output voltage and current. Even so, these values remain closely tied to engine speed and not to electrical loads.

More sophisticated circuits use a solid-state regulator to synchronize charging current and voltage with battery requirements. The regulator caps voltage output at about 14.7 V and responds to low battery terminal voltage with more current.

The usual practice is to encapsulate the regulator with the rectifier. Look for a potted "black box" or a finned aluminum can, mounted under or on the engine shroud. All of these units share engine ground with the stator and battery. Hold-down bolts must be secure and mating surfaces clean.

Most regulator/rectifiers have three wires going to them, as shown in Figures 6-2 and 6-15. Two of these wires carry ac from the stator and one conveys B+ voltage to the battery. Wires that supply B+ are often, but not always, color-coded red.

Twenty and 30A systems can include a bucking coil (a kind of electrical brake) to limit output. The presence of such a coil is signaled by one (or sometimes two) additional wires from the stator to regulator-rectifier.

Use a high-impedance meter, preferably digital, for voltage checks. Identify circuits before testing, with particular attention to the magneto primary, which is often integrated into the regulator-rectifier connector. That circuit carries some 300 Vac.

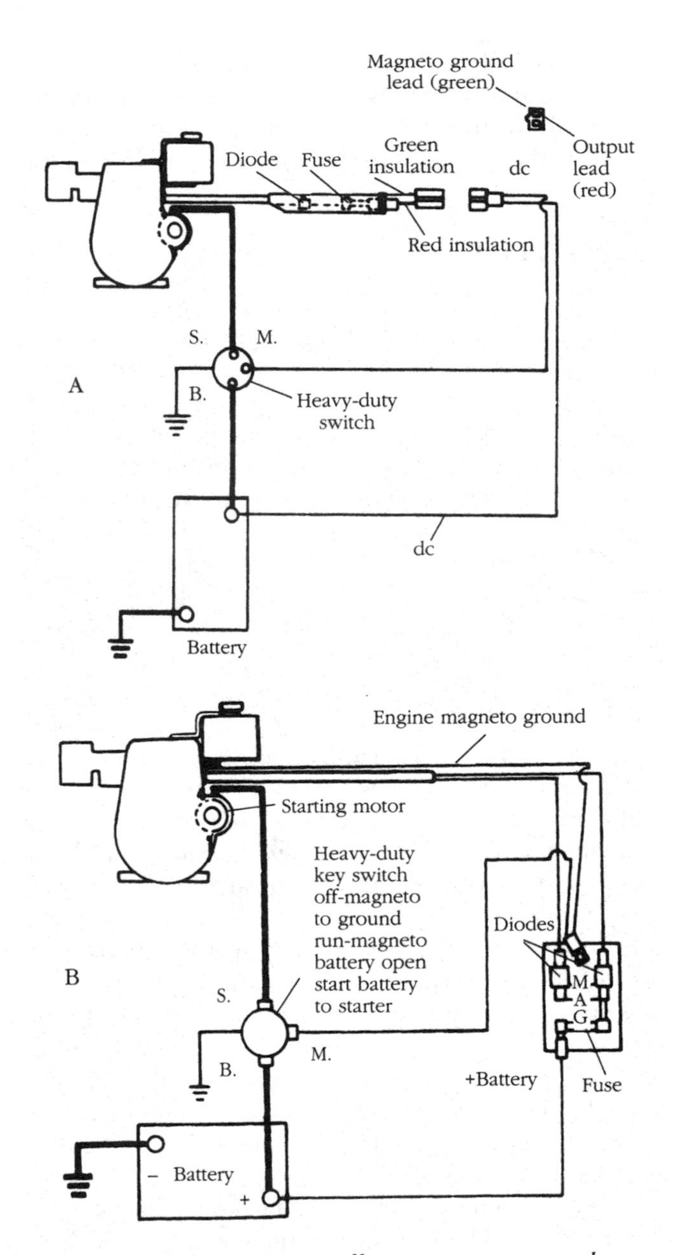

FIGURE 6-14. *Tecumseh 3A systems illustrate two approaches to rectification. The single-diode, half-wave rectifier, located in wiring harness, passes half of stator output to battery (A). The two-diode, full-wave rectifier utilizes all of stator output, doubling the charge rate (B). In the event of overcharging, one diode can be removed.*

Do not:

- Reverse polarity—reversed battery or jumper cable connections will ruin the regulator/rectifier on all but the handful of systems that incorporate a blocking diode.
- Introduce stray voltages—disconnect the B+ rectifier-to-battery lead before charging the battery or arc welding.
- Create direct shorts—do not ground any wire or touch ac output leads together.
- Operate the system without a battery—when open-circuit, unregulated ac output tests are permitted, make them quickly at the lowest possible engine rpm/voltage needed to prove the stator.
- Run the engine without the shroud in place—if necessary, route test leads outside of the shroud.

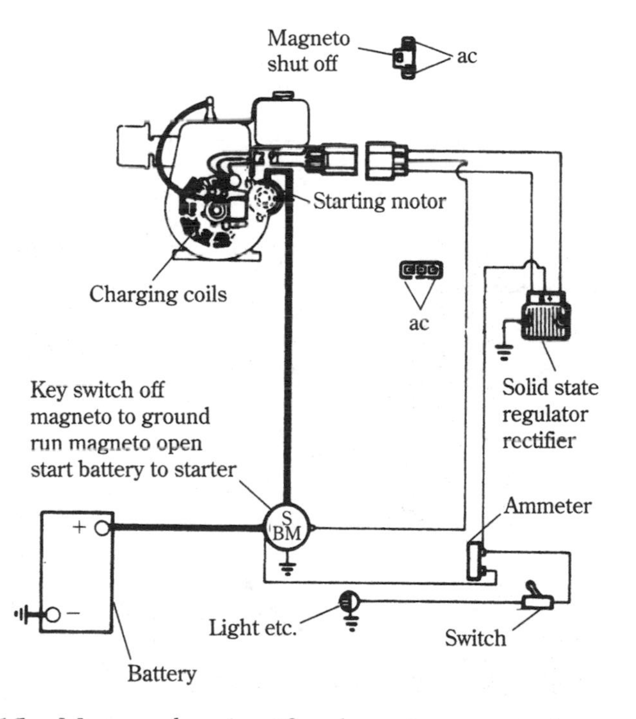

FIGURE 6-15. *Most regulator/rectifiers have three terminals—ac, ac, B+—and ground to the engine through the hold-down bolts.*

Figure 6-16 presents a standard troubleshooting format used by many small engine mechanics. It applies to all unregulated systems and to more than 90 percent of systems with a regulator or regulator-rectifier. There are exceptions: Certain regulators and regulator/rectifiers do not tolerate hot (engine running) disconnects. These components will be damaged by attempts to measure open-circuit ac voltage.

An extensive inquiry has uncovered two of these maverick systems; there are almost certainly others among the thousands of models and types of small engines sold in the United States (Kawasaki, for example, lists more than 300 distinct models).

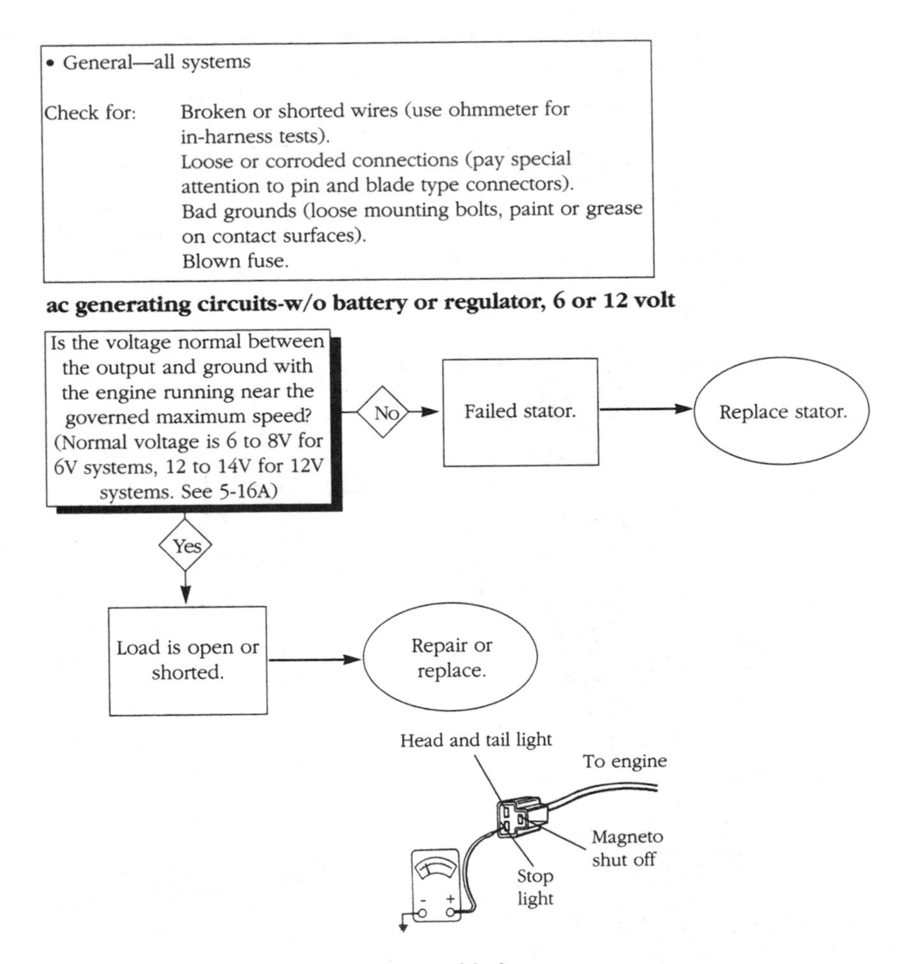

FIGURE 6-16. *Charging system troubleshooting.*

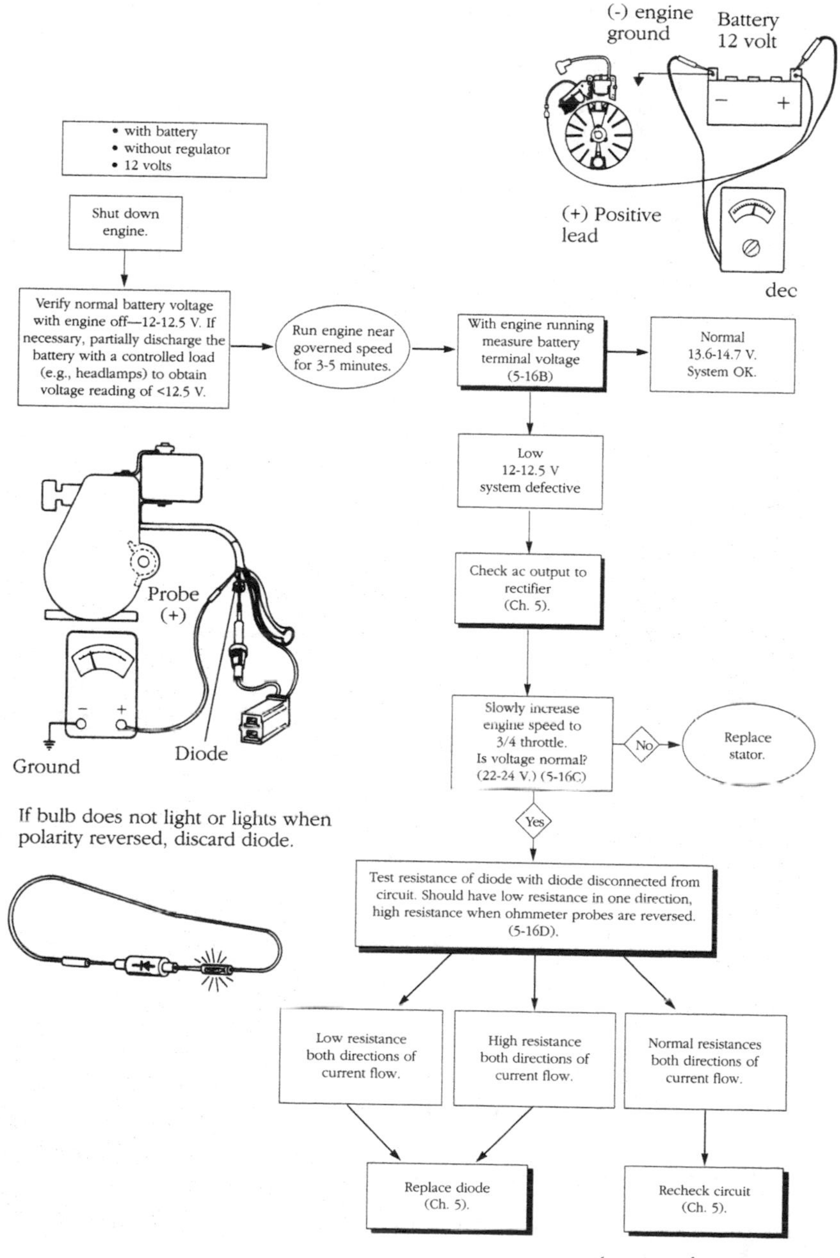

(continued on next page)

FIGURE 6-16. *(Continued)*

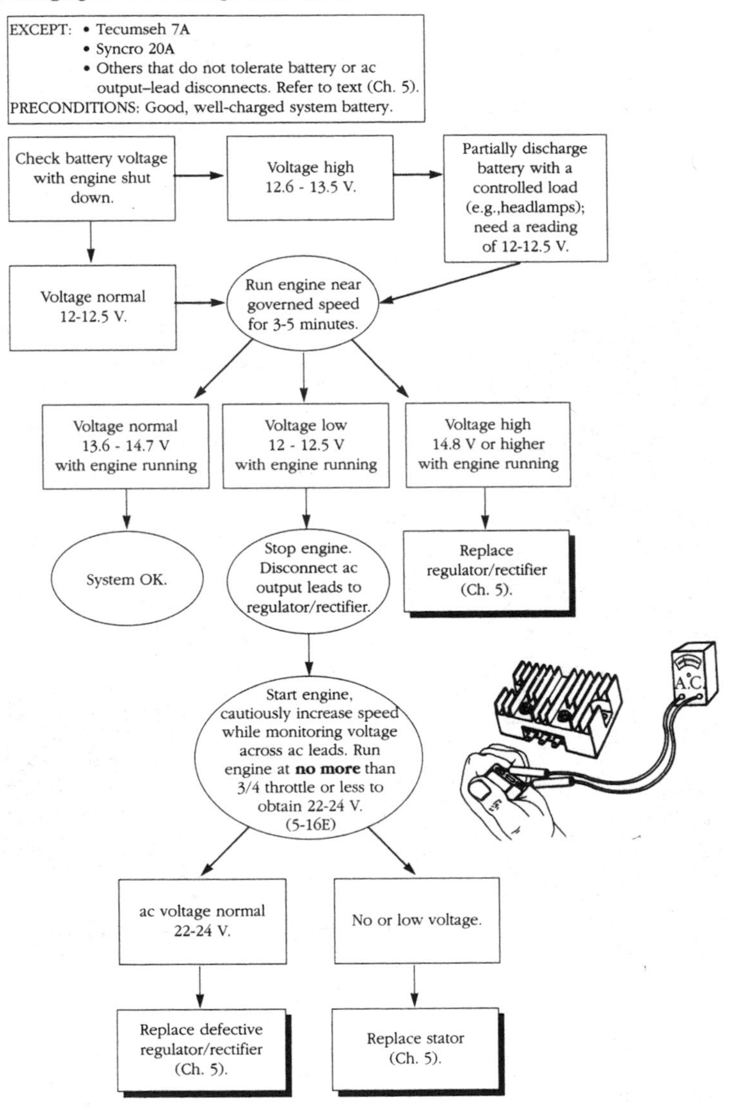

FIGURE 6-16. *(Continued)*

Caution: On any system you are not familiar with, contact a factory-trained mechanic or a manufacturer's tech rep before making hot disconnects.

The two known mavericks are the Tecumseh-supplied 7A system and the Synchro 20A system with separate regulator and rectifier. The Tecumseh 7A system, found on some 3- to 10-hp side valve engines and the overhead valve OMV 120 uses any of three under-shroud regulator-rectifiers shown in Figure 6-17; the caption describes the ac voltage test procedure. Synchro regulators and rectifiers are clearly labeled with their manufacturer's name. Bring these systems to a dealer for service.

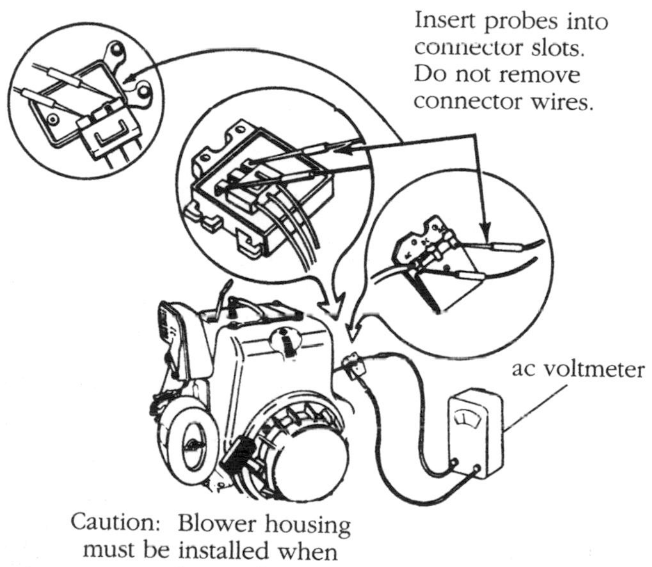

FIGURE 6-17. *Tecumseh 7A system cannot tolerate open-circuit ac voltage measurements. Test ac output as shown, with regulator/rectifier electrically connected and the engine cooling shroud in place. (Unlike other rectifier/regulators, these units do not require an engine ground.) Minimum acceptable stator performance is: 16 Vdc at 2500 rpm; 19 Vac at 3000 rpm; 21 Vac at 3300 rpm.*

7

Engine mechanical

Chapter 2 describes quick checks for compression, bearing side play, and crankshaft straightness that should be made before an engine is torn down. Table 7-1 lists the more common two- and four-cycle maladies, together with their probable causes. Figure 7-1 describes why four-cycle engines develop a thirst for oil, and Figure 7-2 explains where the power goes.

Repairs to two-stroke engines are generally confined to piston-ring and crankshaft-seal replacement. More comprehensive repairs are impractical, at least for discount-house throwaways (Fig. 7-3).

Cylinder head

Four-cycle engines employ demountable cylinder heads sealed with composition gaskets and secured by cap screws.

Warning: When dealing with vintage engines, treat the gaskets as toxic material. Asbestos was phased out during the 1970s, but replacement gaskets may date from the time when this material was used.

Remove carbon deposits from the combustion chamber with a dull knife and a wire wheel. Try not to gouge the aluminum, especially the gasket surfaces.

Check head distortion with the aid of a piece of plate glass. If a 0.003-in. feeler gauge can be inserted between the bolt holes, the gasket surface should be refinished (Fig. 7-4). Tape a sheet of medium-grit wet-or-dry emery paper to the plate glass and, applying pressure to the center of the casting, grind in a figure-8 pattern. Oil makes the work go faster. Stop when the surface takes on a uniform sheen.

TABLE 7-1. Engine-related malfunctions (assuming ignition, fuel, and starting systems are functional).

Symptom	Probable causes
Crankshaft locked	Jammed starter drive Hydraulic lock—oil or raw fuel in chamber Rust-bound rings (cast-iron bores only) Bent crankshaft Parted connecting rod Broken camshaft
Crankshaft drags when turned by hand	Bent crankshaft Lubrication failure, associated with cylinder bore and/or connecting rod
Crankshaft alternately binds and releases during cranking (rewind or electric starter)	Bent crankshaft Incorrect valve timing Loose blade/blade adapter (rotary lawnmower) Loose, misaligned flywheel
No or weak cylinder compression	Blown head gasket Leaking valves Worn cylinder bore/piston/piston rings Broken rings Holed piston Parted connecting rod Incorrect valve timing
No or imperceptible crankcase compression (two-cycle)	Leaking crankcase seals Leaking crankcase gaskets Failed reed valve (engines so equipped)
Rough, erratic idle	Stuck breather valve Leaking valves

Symptom	Probable causes
Misfire, stumble under load	Improper valve clearance Weak valve springs Leaking carburetor flange gasket Leaking crankcase seals (two-cycle)
Loss of power	Loss of compression Leaking valves Incorrect valve timing Restricted exhaust ports/muffler (two-cycle) Leaking crankcase seals (two-cycle)
Excessive oil consumption (four-cycle)	Faulty breather Worn valve guides Worn or glazed cylinder bore Worn piston rings/ring grooves Worn piston/cylinder bore Clogged oil-drain holes in piston Leaking oil seals
Engine knocks	Carbon buildup in combustion chamber Loose or worn connecting rod Loose flywheel Worn cylinder bore/piston Worn main bearings Worn piston pin Excessive crankshaft endplay Excessive camshaft endplay Piston reversed (engines with offset piston pins) Loose PTO adapter
Excessive vibration	Loose or broken engine mounts Bent crankshaft

(continued on next page)

TABLE 7-1. *(Continued)*

Symptom	Probable causes
Oil leaks at crankshaft seals	Hardened or worn seals Scored crankshaft Bent crankshaft Worn main bearings Scored oil seal bore allowing oil to leak around seal outside diameter Seal tilted in bore Seal seated too deeply in bore blocking oil return hole Breather valve stuck closed
Crankcase breather passes oil (four-cycle)	Leaking gasket Dirty or failed breather Clogged drain hole in breather box Piston ring gaps aligned Leaking crankshaft oil seals Valve cover gasket leaking (overhead valve engines) Worn rings/cylinder/piston

Install the new gasket, lubricate the bolts with 30-weight motor oil, and torque in three equal increments to specification. Four-bolt heads tighten in an X pattern. Others are torqued as one would iron a shirt, that is from the center outward. But shown in Figure 7-5, this general rule does not always hold. Consult the factory manual for the engine in question.

Valves

Either of the spring compressors shown in Figure 7-6 can be used to extract and install side (block-mounted) valves. Rotate the crankshaft to seat the valves, insert the tool under the valve collar, compress the spring, and withdraw the locks.

Lock installation goes easier with a magnetic insertion tool, such as a Snap-on CF 771. When properly seated, the locks are swallowed by the collar and no longer visible.

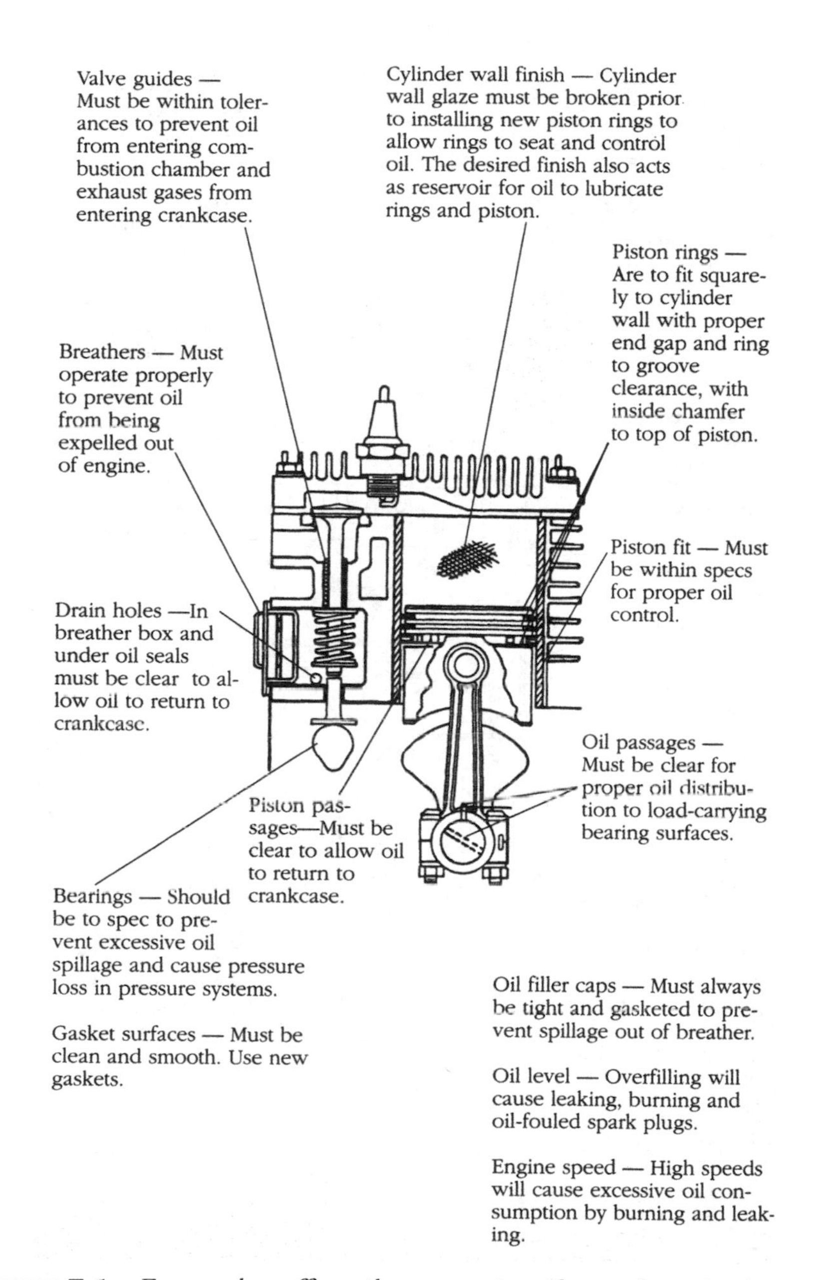

FIGURE 7-1. *Factors that affect oil consumption (four-cycle engines).*

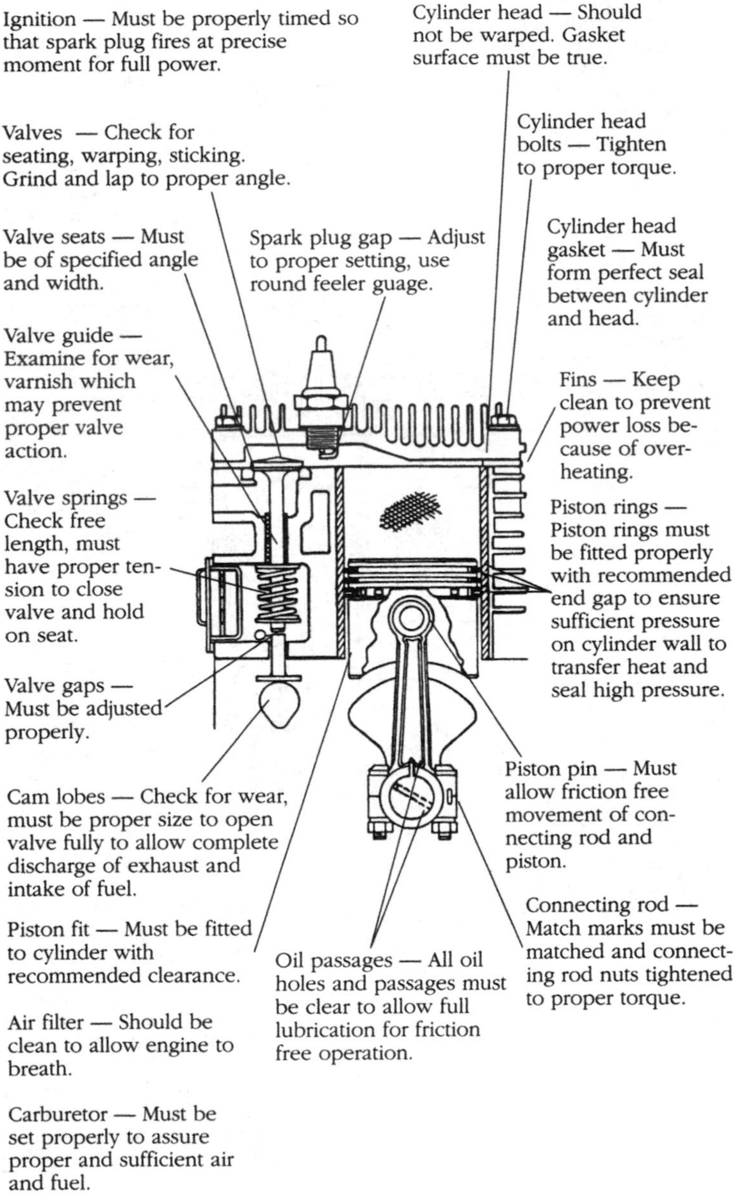

Ignition — Must be properly timed so that spark plug fires at precise moment for full power.

Cylinder head — Should not be warped. Gasket surface must be true.

Valves — Check for seating, warping, sticking. Grind and lap to proper angle.

Cylinder head bolts — Tighten to proper torque.

Valve seats — Must be of specified angle and width.

Spark plug gap — Adjust to proper setting, use round feeler guage.

Cylinder head gasket — Must form perfect seal between cylinder and head.

Valve guide — Examine for wear, varnish which may prevent proper valve action.

Fins — Keep clean to prevent power loss because of overheating.

Valve springs — Check free length, must have proper tension to close valve and hold on seat.

Piston rings — Piston rings must be fitted properly with recommended end gap to ensure sufficient pressure on cylinder wall to transfer heat and seal high pressure.

Valve gaps — Must be adjusted properly.

Piston pin — Must allow friction free movement of connecting rod and piston.

Cam lobes — Check for wear, must be proper size to open valve fully to allow complete discharge of exhaust and intake of fuel.

Piston fit — Must be fitted to cylinder with recommended clearance.

Oil passages — All oil holes and passages must be clear to allow full lubrication for friction free operation.

Connecting rod — Match marks must be matched and connecting rod nuts tightened to proper torque.

Air filter — Should be clean to allow engine to breath.

Carburetor — Must be set properly to assure proper and sufficient air and fuel.

FIGURE 7-2. *Factors that affect power output (four-cycle engines).*

FIGURE 7-3. *The Ryobi in the lower part of the photo was repaired with a connecting rod and piston cannibalized from the second engine. New parts would have been prohibitive.* Robert Shelby

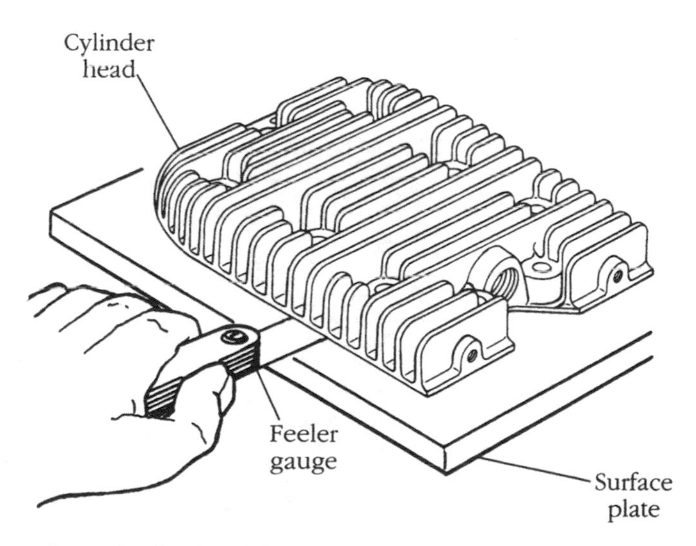

FIGURE 7-4. *Cylinder head flatness should be checked to assure gasket integrity. A piece of plate glass can be substituted for the surface plate shown.*

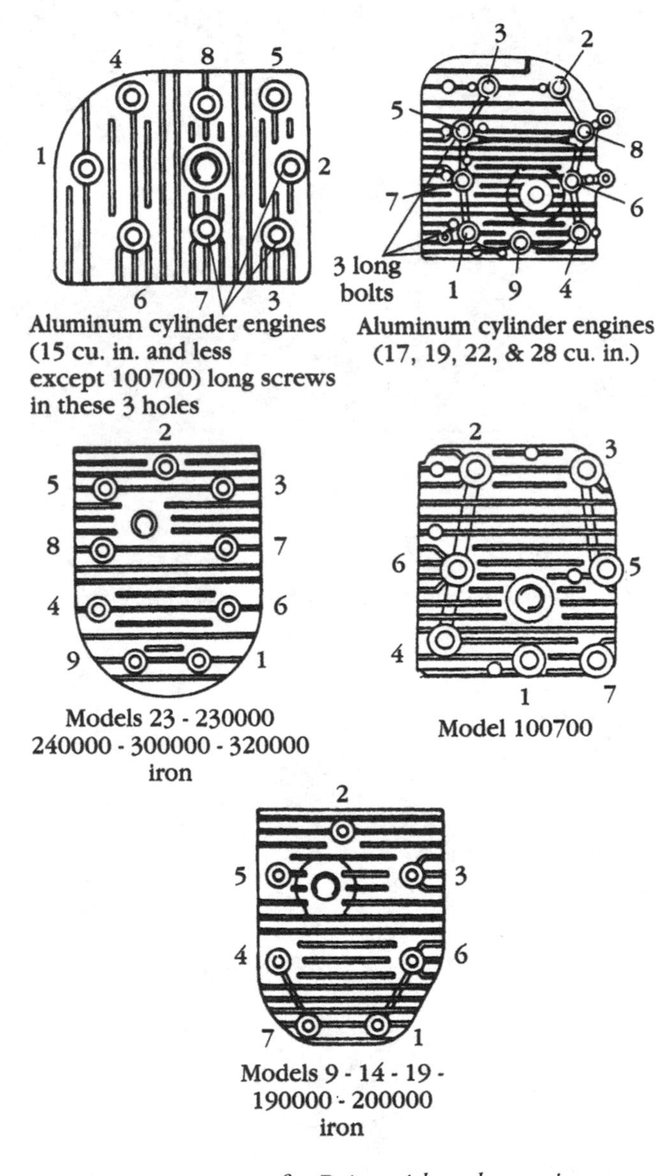

FIGURE 7-5. *Torque sequence for Briggs side-valve engines.*

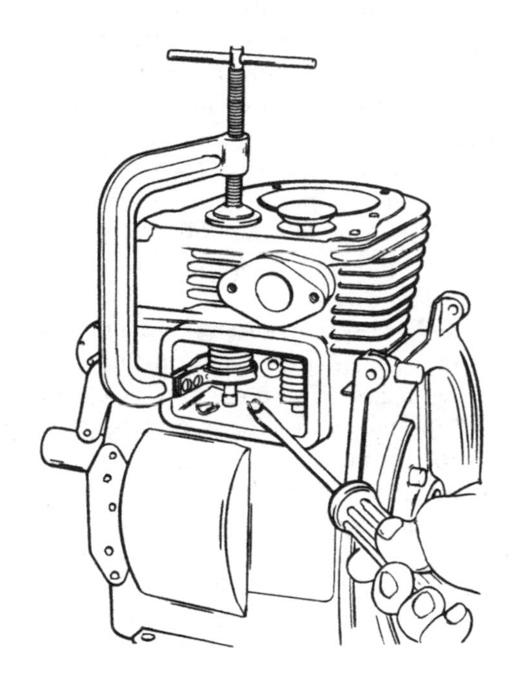

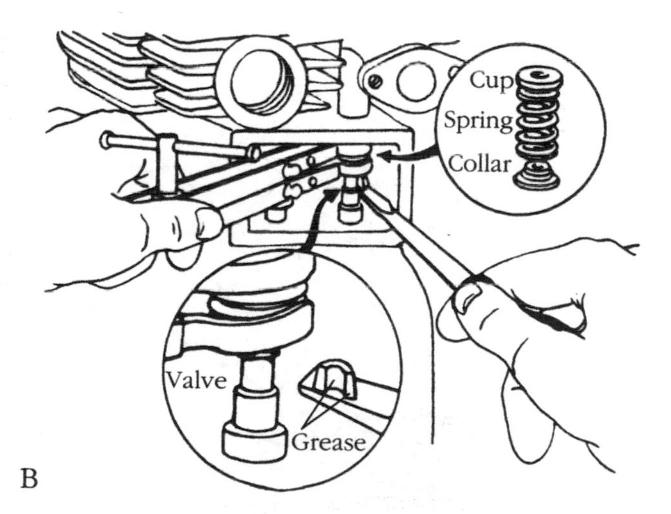

FIGURE 7-6. *Use a clamp (A) or bridge-type (B) spring compressor to remove and install valves mounted in the block. The former tool is available from Kohler and the latter from Briggs & Stratton. Note how split valve locks are spooned into place with a grease-coated screwdriver.*

Overhead valves have better accessibility. Detach the cylinder head and support the casting on a wood block to avoid scaring the gasket surface. Some valve springs compress with finger pressure. Japanese and Tecumseh springs are stouter and require a compressor tool, such as shown in Figure 7-7. In an emergency, one can disengage the locks with impact. Place a soft wood block under the valve face and a large socket wrench over the valve collar. A hammer blow on the socket compresses the spring and pops the keepers off. Assembly without the proper compressor tool can be accomplished by squeezing the springs in a vise and wrapping them with fine-gauge wire. The wire is retrieved after installation.

Exhaust and intake valve springs often—but not always—interchange. When they do not, the heavier spring goes on the exhaust side. Some engines employ springs with closely wound damper coils on the stationary side of the spring (Fig. 7-8). That is, the damper coils go on the end furthest from the actuating mechanism.

Valve springs should stand flat, conform to manufacturer's specifications for freestanding height, and exhibit no signs of coil binding or stress pitting.

Worn valve faces and seats should be turned over to a dealer or automotive machinist for servicing. Figure 7-9 shows a commercial grinder in use. While most small-engine valves are cut at 45°, Onan likes 44° and Briggs sometimes employs 30° on the intakes and 45° on the exhaust. Valve work goes nowhere without factory documentation (Fig. 7-10). Let the shop know that you have access to the documentation and will inspect their work to see that it conforms to it.

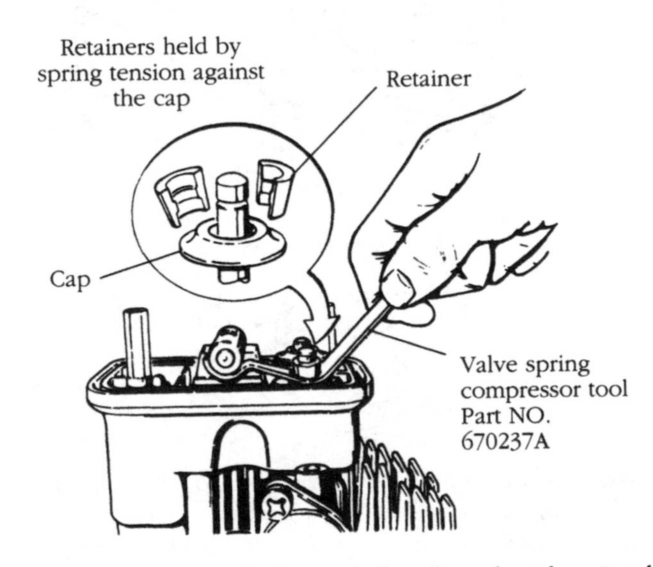

FIGURE 7-7. *Overhead-valve keepers can be released with a simple spring compressor or, as explained in the text, shocked loose.*

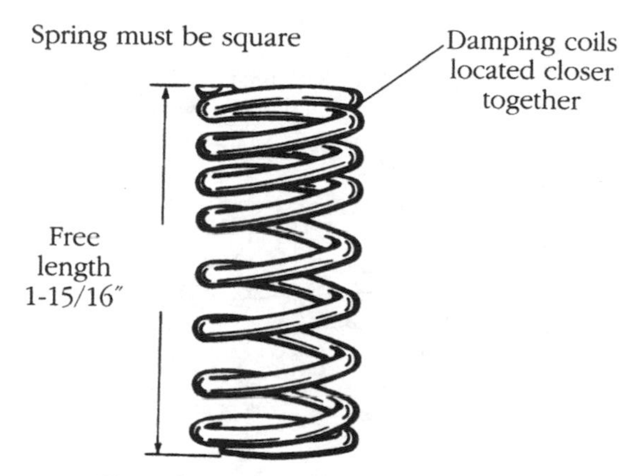

FIGURE 7-8. *Install variable-rate springs with the damper coils on the stationary ends, furthest from the actuating mechanism. Springs should be replaced as part of every overhaul, especially on ohv engines where spring failure can result in a swallowed valve.*

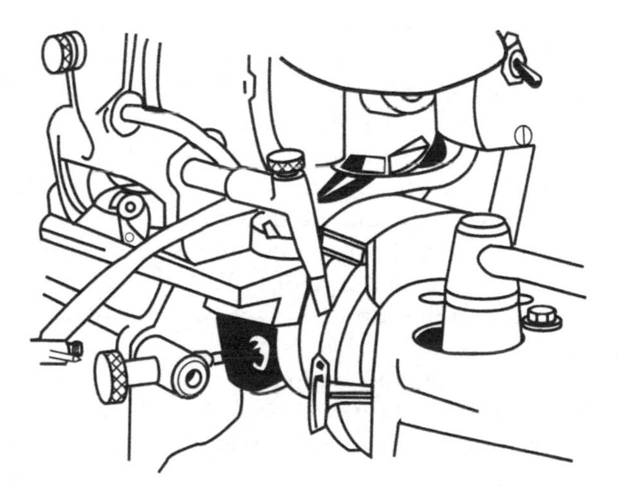

FIGURE 7-9. *A high-speed valve grinder.*

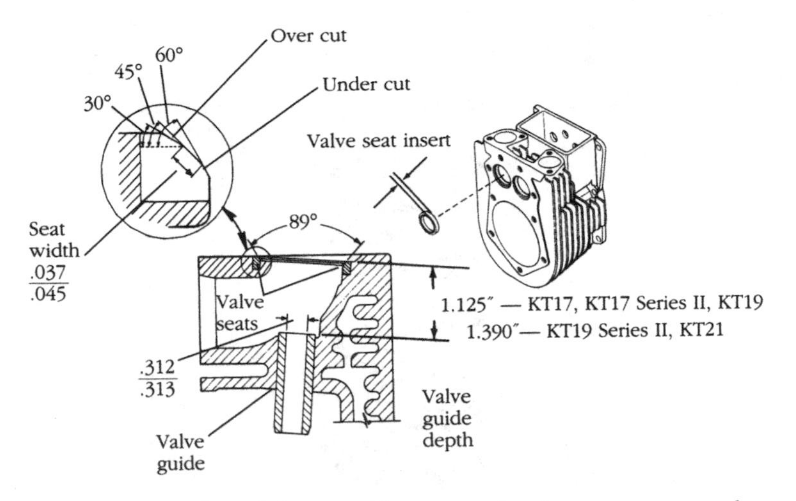

FIGURE 7-10. *An idea of the crucial nature of valve geometry can be seen from this Kohler-supplied drawing.*

Valve guides

As a rule, the wear limit for valve guides is 0.004 in., a figure difficult to detect without a set of plug gauges. If a valve exhibits perceptible wobble when wide open, the guide probably needs replacement. Some engine makers supply valves with oversized stems, so that the original guide can be reamed oversize. Others would have you replace and, if necessary, ream the guides.

Normally this is work for a machinist. However, a patient mechanic can usually pull it off. Begin by measuring the installed depth of the original guides, that is, the distance from the top of the guides to the valve seats, as called out in Figure 7-10. Side-valve guides knock out and install from above (Fig. 7-11). You may have to shatter the old guides with a punch to retrieve them from the valve chamber.

Fabricate a pilot driver with a reduced diameter on one end that exactly matches the guide ID. Drive the new guide to installed depth and test with a valve. If the valve binds, ream the guide 0.0015-0.0020 in. larger than the stem diameter. Because of the odd sizes involved, you will need to use an adjustable reamer.

Overhead-valve heads should be heated in oil prior to guide service (Fig. 7-12). While this complicates the job, the reduced installation force usually eliminates the need for a special reamer.

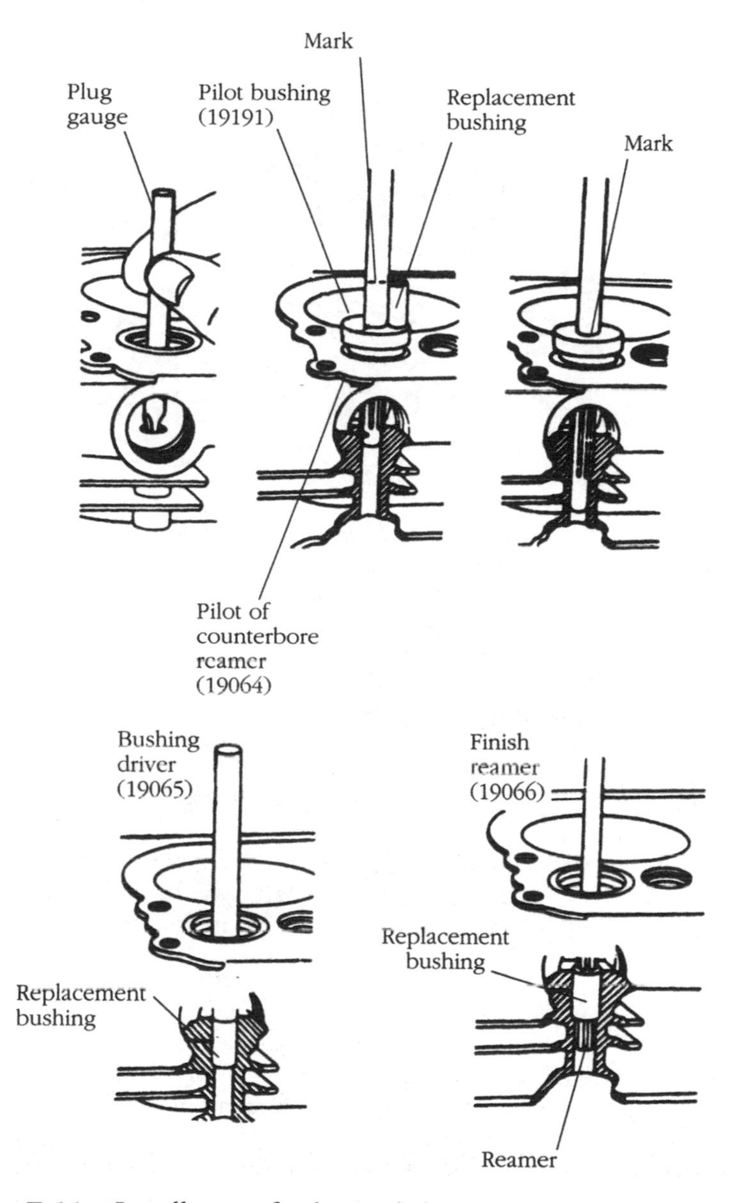

FIGURE 7-11. *Installation of valve guide bushings goes much easier if you have the correct tools.*

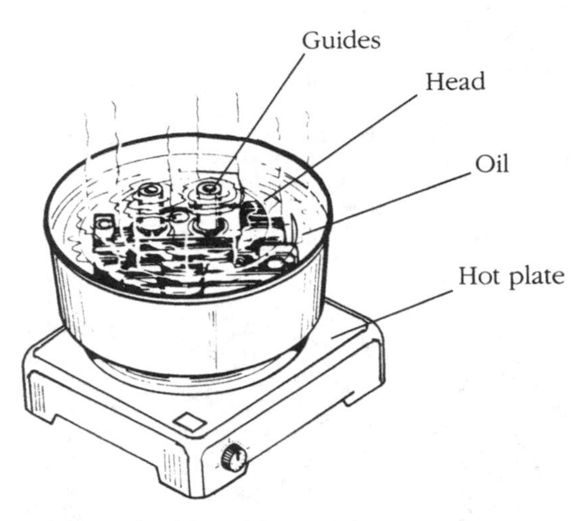

FIGURE 7-12. *Aluminum ohv heads do not take kindly to brute-force methods of valve guide extraction and installation. Heat the head in oil while supporting it off the bottom of the container.*

Valve seats

Loose valve seats can sometimes be repaired by peening, although don't bet on it (Fig. 7-13). Nor should worn or cracked seats be removed with a punch from below or pried out with an extractor (Fig. 7-14), and new ones hammered home with a valve as the pilot. For a reliable repair, seat recesses should be machined to fit the replacement part (which will be slightly oversized) and the seats chilled to reduce installation force. Dry ice and alcohol produce the lowest temperatures.

Valve lash adjustment

Most side-valve engines have nonadjustable tappets, which means that metal lost to the valve face or seat must be compensated for by grinding the tip of the valve stem.

Install the valve without the spring, turn the crankshaft until the valve rides on the heel of the cam, and measure the clearance with a feeler gauge (Fig. 7-15). Carefully grind the stem, just "kissing" the wheel, to obtain the specified clearance (typically 0.006-0.008 in. for the intake and 0.010-0.013 in. for the exhaust). Take off too much and the associated valve or seat will have to be reground. Finish by breaking the sharp edges with a stone.

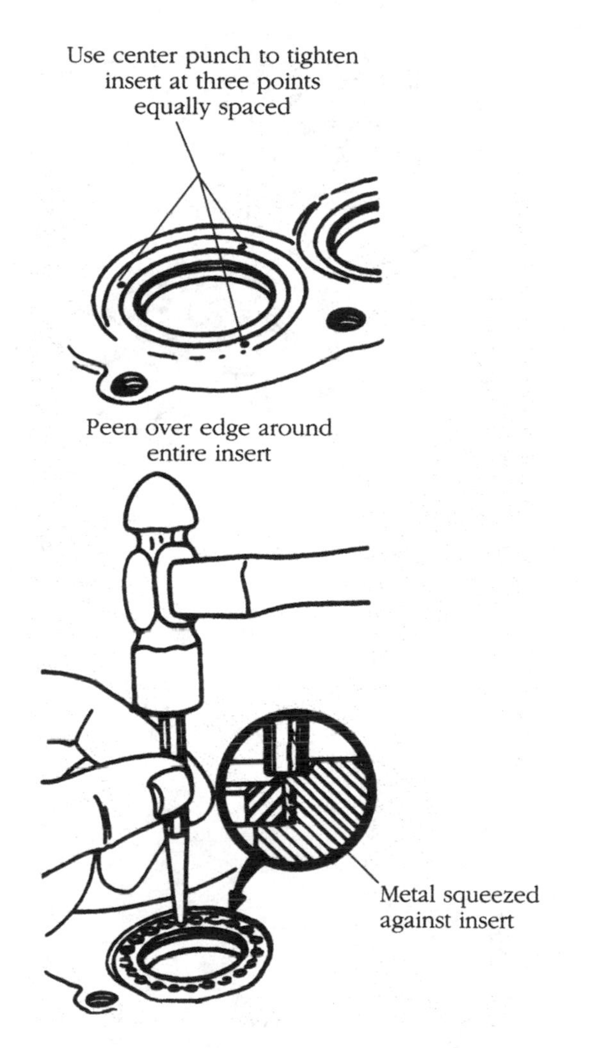

FIGURE 7-13. *Some mechanics attempt to repair loose valve seats by peening. That rarely, if ever, works. Peening is also used as insurance when new seats are installed. A better approach is to have the machinist recess the new seat about 0.030 in. below the surrounding metal. Then, using a flat-tipped punch, roll the metal over the edge of the seat.*

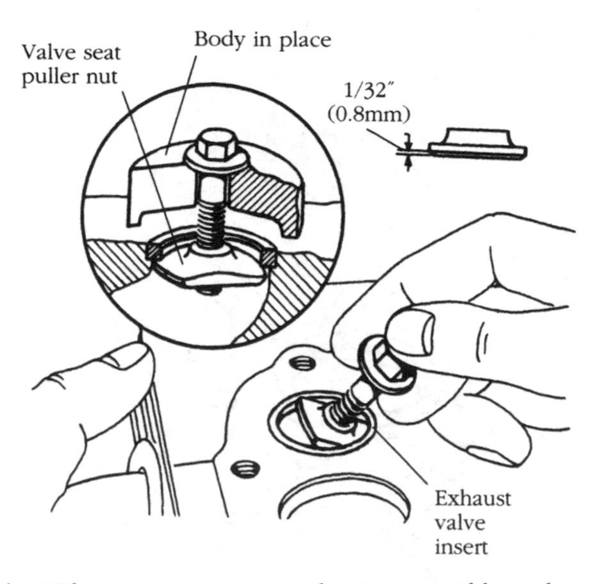

FIGURE 7-14. *When port geometry makes it impossible to drive the seats out from below with a punch, a seat puller fabricated from an old valve can be used. However, neither of these methods is optimal. To eliminate damage to the seat bores, have a machinist cut the old seats out.*

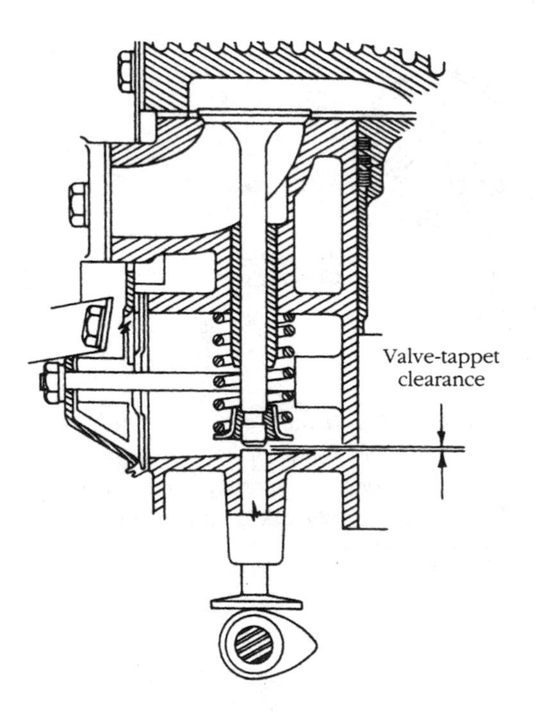

FIGURE 7-15. *Valve lash on side-valve engines is measured between the end of the valve stem and the tappet, with the tappet on the cam base circle.*

Some of the better side-valve and all overhead-valve engines have provisions for valve-lash adjustment. Lash for ohv engines appears as the clearance between the rocker arm and valve stem. Adjustment screws for engines with shaft-supported rocker arms bear against the pushrods (Fig. 7-16). Other ohv engines use stamped-steel rockers that pivot on fulcrum nuts. The fulcrum nuts, secured to their studs by setscrews or locknuts, control lash by varying the height of the rocker arms (Fig. 7-17).

To adjust lash on ohv engines, rotate the crankshaft to bring the associated tappet on the heel of the camshaft and loosen the locknut or screw. Move the adjustment screw or fulcrum nut as necessary to achieve the required clearance. Tighten the lock and check the adjustment, which will have drifted a few thousandths.

Push rods

Inspect push rods for wear caused by contact with the guide plates and for straightness. It is possible to salvage bent rods with judicious vise work, but the practice is an expedient, resorted to when replacement parts cannot be had.

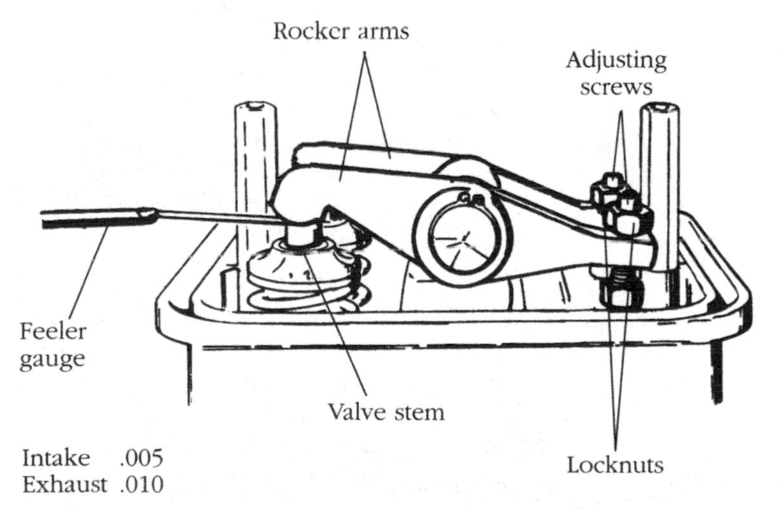

FIGURE 7-16. *Lash for overhead valves is measured between the end of the valve stem and rocker arm. Shaft-mounted rockers carry the adjustment screws.*

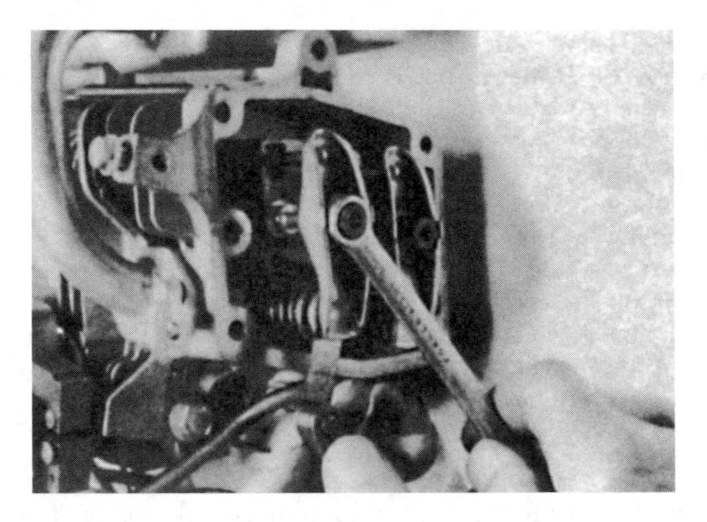

FIGURE 7-17. *Pressed-steel rockers pivot on adjustable fulcrum nuts, secured by set screws or lock nuts.*

Breathers

Four-cycle engines include a crankcase breather connected by a flexible line to the carburetor intake (Fig. 7-18). The filter element, in conjunction with a baffle, separates liquid oil from blow-by gases, which then recycle through the carburetor. The reed valve maintains a slight negative pressure in the crankcase to reduce seepage past gaskets and crankshaft seals. If the filter clogs or the valve becomes inoperative, the engine pumps oil like a mosquito fogger.

Reed valves

Many two-cycle engines use a functionally similar device in the form of a reed valve between the carburetor and crankcase (Fig. 7-19). The reed assembly acts as a check valve to contain the air-fuel mixture in the crankcase. Contact surfaces should be dead flat, and valve petals should either rest lightly on their seats or stand off by no more than a few thousandths of an inch.

Tecumseh also incorporates a reed-type compression release in some of its two-strokes. While these devices are rarely encountered, the engineering is worth showing (Fig. 7-20).

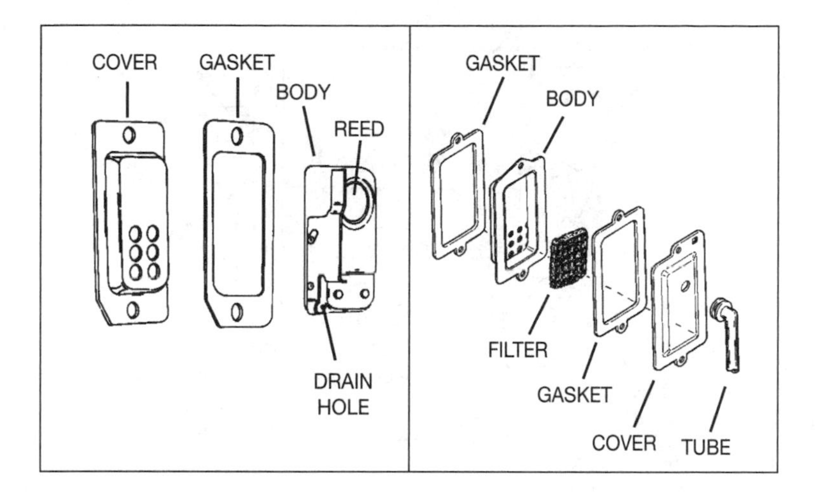

FIGURE 7-18. *Various Tecumseh crankcase breathers.*

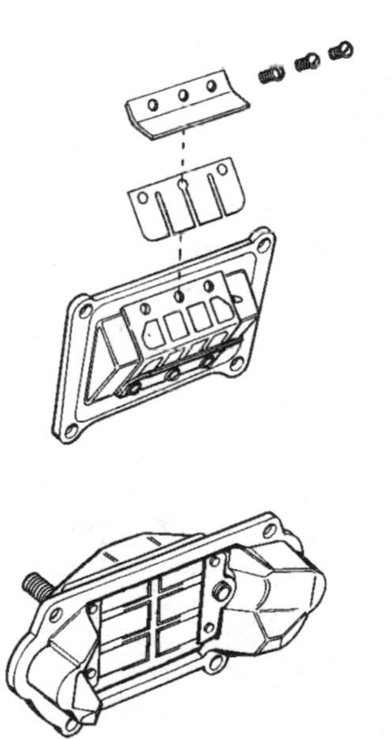

FIGURE 7-19. *Reed intake valves should stand off no more than 0.010 in. from their seats.* Tecumseh Products Co.

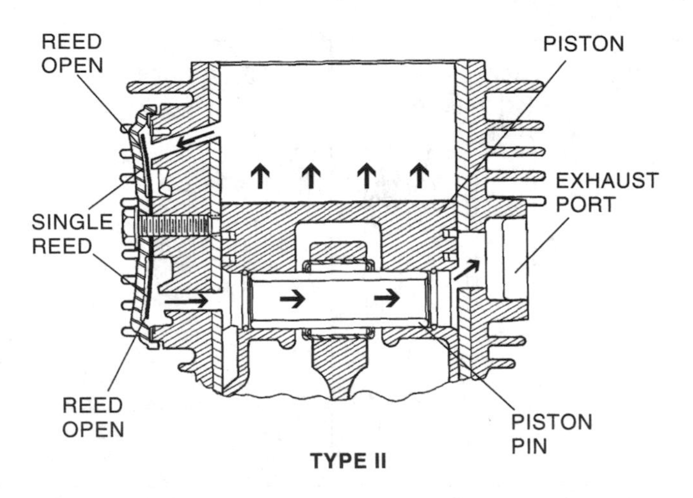

FIGURE 7-20. *Tecumseh two-stroke compression release opens during cranking to bleed pressure through the piston pin and out the exhaust port.*

Pistons and rings

Disengage the flange or side cover as described in the caption to Figure 7-21. Two-piece connecting rods have their caps secured by bolts or studs. To avoid an assembly error, make note of the orientation of the rod-and-piston assembly relative to the camshaft or some other prominent feature. Loosen rod nuts in two or three steps, and remove the rod cap. Match marks on rod cap and shank must be aligned upon assembly. Drive the piston and attached rod shank out of the top of the bore with a hammer handle or wooden dowel (Fig. 7-22).

Figure 7-23 shows the architecture of a typical two-stroke engine. In this example, the crankcase parting line passes through the center of the cylinder bore. Each half of the crankcase carries a press-fitted ball- or roller-type main bearing and oil seal. String-trimmer and other inexpensive engines often get by with a single main bearing.

Note: Weed-eater centrifugal clutches may have left-hand threads.

Inspection

Bright, uniformly polished rings are the norm. Rings that stick in their grooves suggest poor maintenance (failure to change oil, dirty cooling fins) or abuse (lugging under load, insufficient power for the application). Broken rings result from improper installation or worn piston grooves.

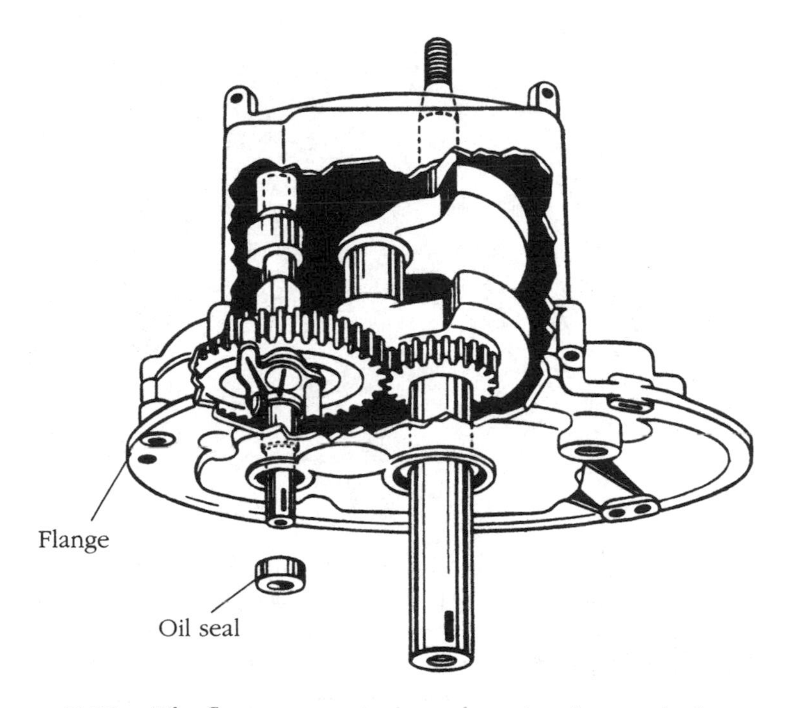

FIGURE 7-21. *The flange on vertical-crank engines locates the lower, or pto, main bearing. Before proceeding with disassembly, remove rust and tool marks from the crankshaft with an emery cloth and a file. Cover the keyways (which are sharp enough to cut the crankshaft seal) with a layer of Scotch tape. Lubricate the crankshaft and remove the flange hold-down cap screws. Position the engine on its side and separate the castings with a rubber mallet. The camshaft (on side- and overhead-head-valve engines) should remain engaged with the flywheel so that timing-mark alignment can be verified. Do not attempt to pry the flange off. The same general procedure holds for side covers on horizontal-shaft engines.*

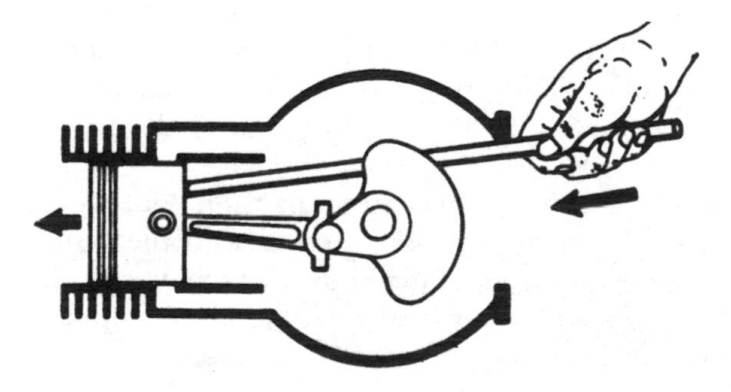

FIGURE 7-22. *Once the rod cap has been detached, use a wooden dowel to drive the piston assembly out.*

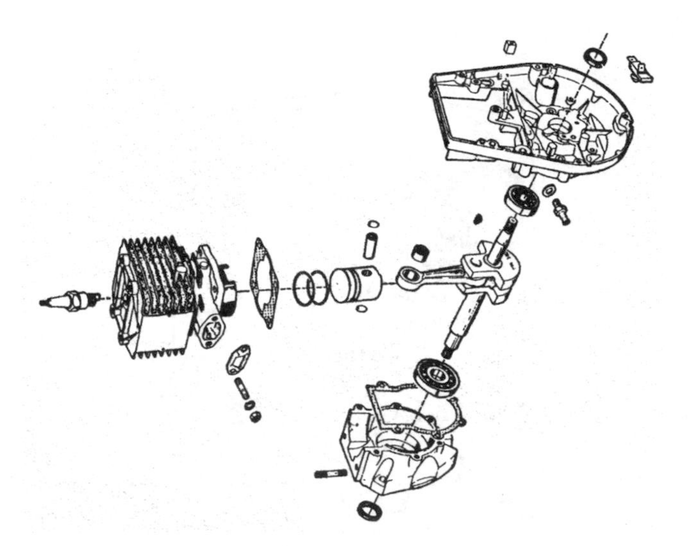

FIGURE 7-23. *Spitting the crankcase on a two-stroke engine with detachable cylinder barrels can pose difficulties. The resistance of main-bearing fits, sealant applied to the crankcase parting lines, and interference fits of locating pins must be overcome before the cases can be separated. Stubborn crankcases can be gently warmed with a propane torch and pried apart with a hammer handle inserted into the cylinder-barrel cavity. Exercise extreme care to avoid over-heating or warping the fragile castings.*

Examine the piston skirt for wear on the thrust faces at right angles to the piston pin. Figure 7-24 illustrates abnormal wear patterns produced by bent or twisted connecting rods. Forces that rock the piston can also drive the piston pin past its locks and into collision with the cylinder bore.

Deep scratches on the piston skirt result from chronic overheating that can leave splatters of aluminum welded to the bore. A dull, matted finish means that abrasives have been ingested, usually by way of a leaking air filter. Should this happen, hone the cylinder bore and replace the piston.

Pistons need about 0.002 to 0.003-in. bore clearance for thermal expansion, but wear limits are flexible. Lightly used four-strokes putter on for years with piston-to-bore clearances of 0.006 in. and more. High-revving two-strokes are less tolerant.

Pistons usually taper toward the crown to allow for expansion under thermal load. In addition, four-cycle pistons are cam ground with the thrust faces on the long axis. The piston remains centered in the bore when cold and expands to a full circle as temperatures increase. Two-stroke pistons are machined round to control leakage at startup.

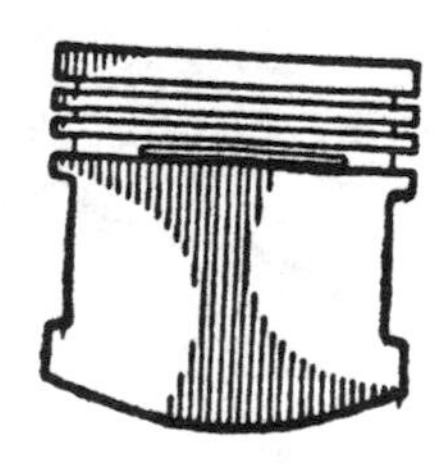

A

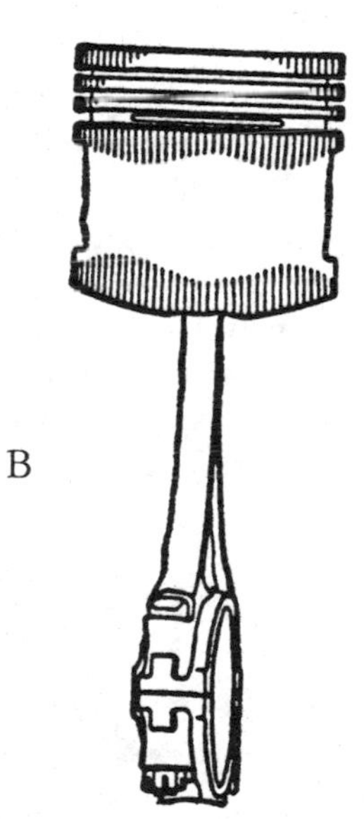

FIGURE 7-24. *As shown by the shaded lines in drawing A, a bent conn rod tilts the piston to create an hourglass-shaped wear pattern. A twisted rod rocks the piston, concentrating wear on the upper and lower edges of the skirt (drawing B).*

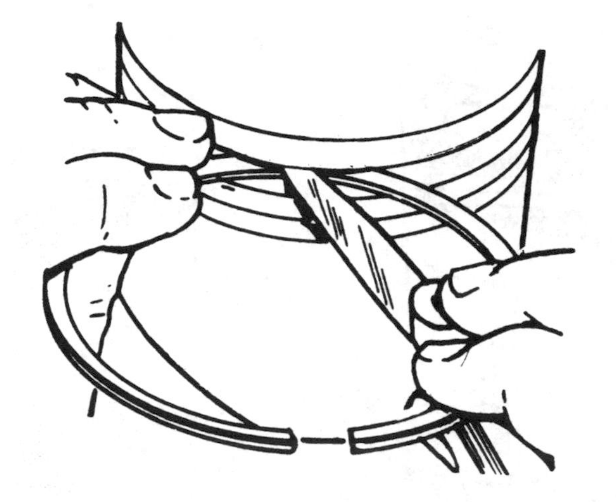

FIGURE 7-25. *Determine ring side clearance using a new ring as a gauge. The upper side of No. 1 groove (shown) takes the worst beating.* Onan

Measurements of piston diameter are made across the thrust faces at 90° to the piston pin. Because of the taper, the measurement must be made at the factory-specified distance from the bottom of the skirt.

The best way to remove carbon from the ring grooves is to farm out the job to an automotive machinist for chemical cleaning. Otherwise, you will need to scrape the grooves with a broken ring mounted in a file handle. (Ring-groove cleaning tools are, in my experience, a waste of money.)

Warning: Piston rings—especially used rings—are razor sharp.

Using a new ring, measure side clearance on both compression-ring grooves as shown in Figure 7-25. (Oil-ring grooves never wear out.) Excessive side clearance, as defined by the manufacturer, allows the ring to twist during stroke reversals (Fig. 7-26).

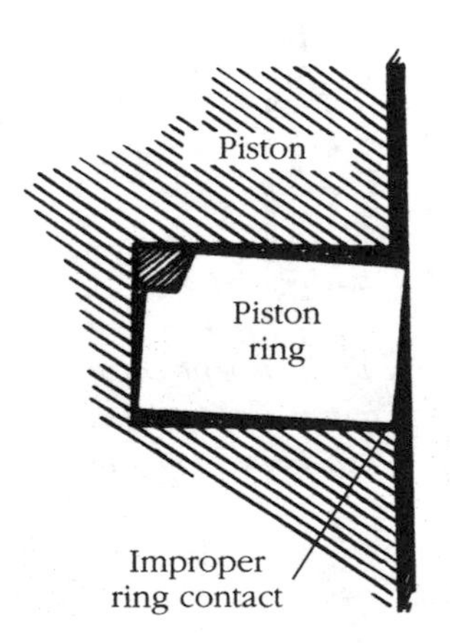

FIGURE 7-26. *A major cause of ring breakage is the twist created by worn ring grooves.*

Piston pins

Four-cycle piston-pin bearing wear is relatively uncommon because of the thrust reversals every second revolution. Compression and expansion strokes bear down on the pin, exhaust and intake strokes drive the pin from below. Two-stroke pins are subject to an almost constant downward thrust that tends to squeeze out the lubricant. In either case, the bearing is considered acceptable if it has no perceptible up-and-down play and if the piston pivots on the pin from its own weight.

Most pistons incorporate a small offset relative to their pins. Consequently, one must install the piston as found. An arrow or other symbol on the crown marks leading edge or indexes with some other reference such as the flywheel.

Remove and discard the locks. New circlips are inexpensive insurance against the pin drifting into contact with the cylinder bore. If the piston and rod assembly are out of the engine, drive or press the pin out, being careful not to score the pin bores. When the connecting rod remains attached to the crankshaft, extract the pin with the tool shown in Figure 7-27 or heat the piston. The safest, and surely the messiest way, to apply heat is to wrap the piston with a rag soaked in hot oil.

Installation is the reverse process. Lubricate the pin and pin bosses with motor oil or assembly lube. Make certain that the pin locks seat around their whole circumferences.

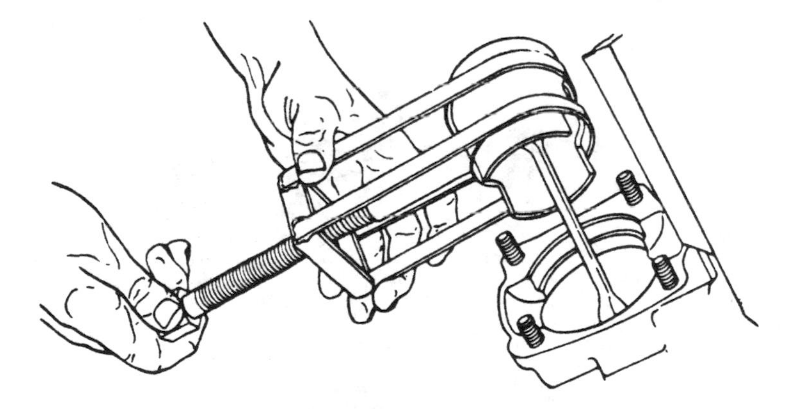

FIGURE 7-27. *A piston-pin extractor can be ordered through motorcycle and snowmobile dealers. A Kohler tool is shown.*

Piston rings

Four-cycle engines usually have three rings. Counting from the top, we have No. 1 compression ring, No. 2 compression (or scraper) ring, and the oil ring. The latter may be cast in one piece or made up of segments. Two-stroke engines are fitted with two identical compression rings, usually fixed in their grooves by pegs. (Were the rings free to rotate, the ends could snag on the ports.)

Careful mechanics check the end gap of each ring. Using the piston crown as a pilot to hold the ring square, insert the ring about midway into the cylinder (Fig. 7-28). Measure the gap with a feeler gauge.

Most manufacturers call for about 0.0015 in. of ring gap per inch of cylinder diameter. Too large a gap suggests that the bore is worn or that the ring is undersized for the application. Too small a gap leads to rapid cylinder wear and shattered rings. Correct by filing the ends square.

Installation

Lay out rings in the order of installation. Make certain that you have correctly identified each ring and each ring's upper side, which should be marked (Fig. 7-29.) The lowest ring goes on first. Using the expander shown in Figure 7-30, spread the ring just enough to slip over the top of the piston and deposit it into its groove. Verify that rings seat into their grooves and that ring ends of two-stroke pistons straddle their pegs.

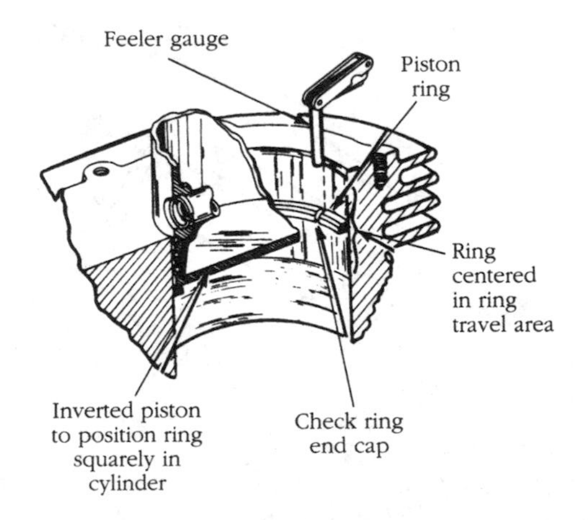

FIGURE 7-28. *Using the piston as a pilot, insert each replacement ring about halfway into the bore and measure its gap. Variations in gap as the ring moves deeper into the bore give some idea of cylinder taper.*

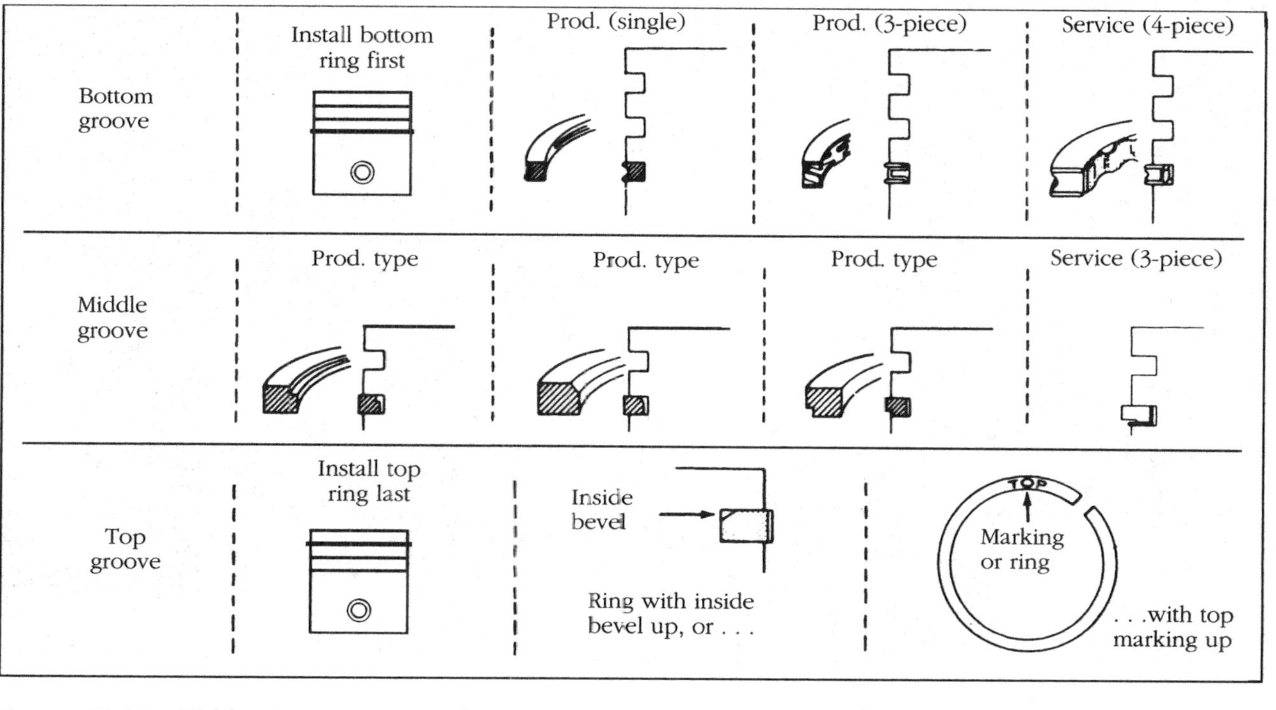

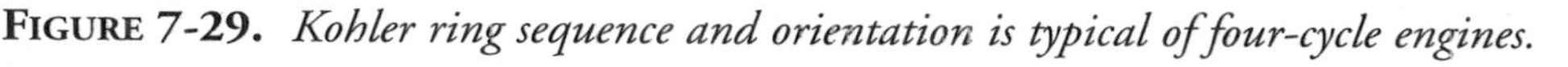

FIGURE 7-29. *Kohler ring sequence and orientation is typical of four-cycle engines.*

In order to contain compression, rotate floating rings to stagger the gaps 120°. On Tecumseh engines with relieved valves (shades of flathead Ford hotrods!), position the ring ends away from the bore undercut (Fig. 7-31).

Integral barrel. Bring the crankshaft to bottom dead center and cover the rod studs (when present) with short pieces of rubber fuel line. Lubricate the cylinder bore, crankpin, rod bearings, pin, and piston with motor oil. Without upsetting the ring-gap stagger, install a compressor tool over the piston (Fig. 7-32). Tighten the band only enough to squeeze the rings flush with the piston diameter.

With the piston oriented as originally found, set the tool hard against the fire deck, and carefully tap the piston into the bore. Do not force matters. If the piston binds, a ring has escaped or the rod has snagged on the crankshaft. Read the "Connecting rod" section before installing the rod cap.

Detachable barrel. Lubricate the cylinder bore, piston pin, and piston ring areas. Support the piston on the crankcase with a wooden fork, as shown in Figures 7-33 and 34. Most factories bevel the lower edge of the bore to facilitate piston entry. Straight-cut bores call for a clamp-type ring compressor (Fig. 7-33).

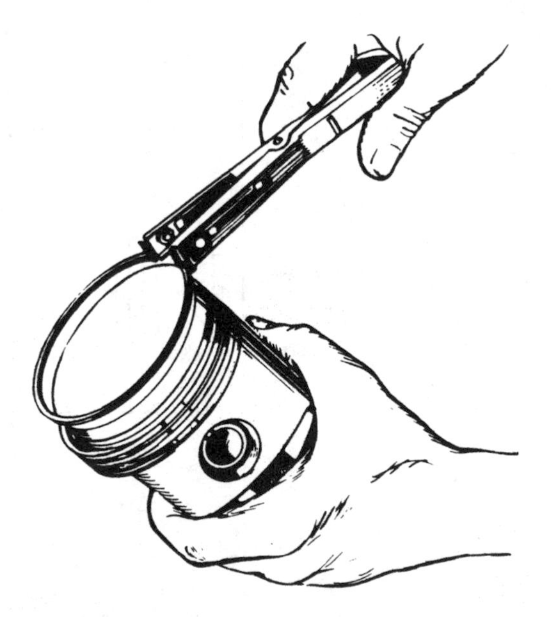

FIGURE 7-30. *Installing a compression ring on an Onan piston with the aid of a ring expander. Expand the rings only enough to slip them over the piston.*

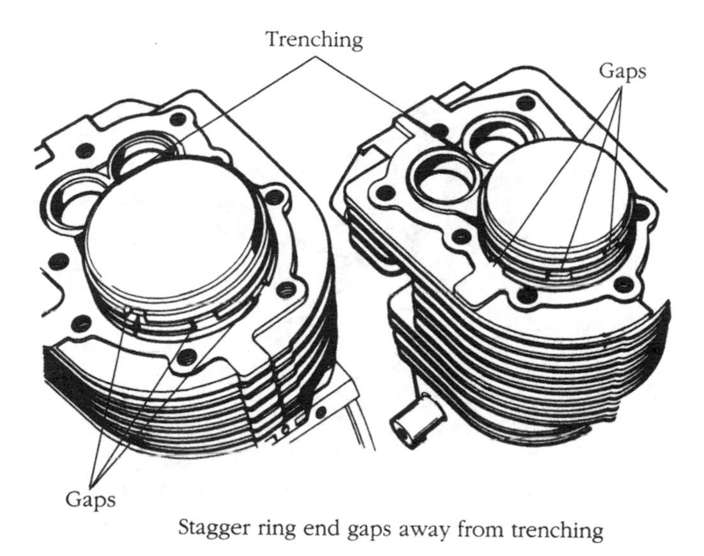

FIGURE 7-31. *Rings for Tecumseh engines with trenched valves install with their ends turned away from the undercut.*

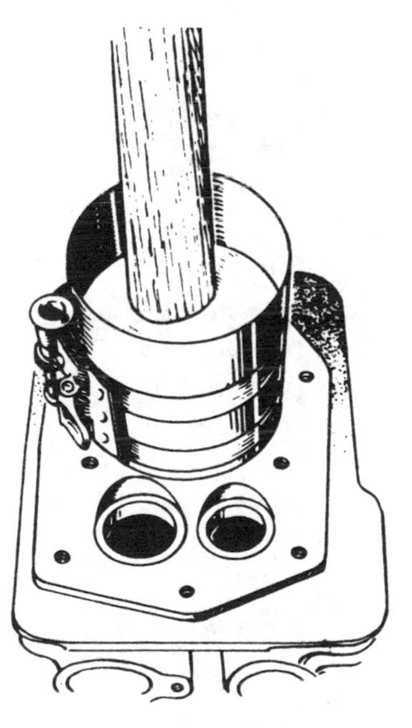

FIGURE 7-32. *A ring compressor sized for small engines is used when the piston installs from the fire deck. Use a hammer handle to gently tap the piston home.*

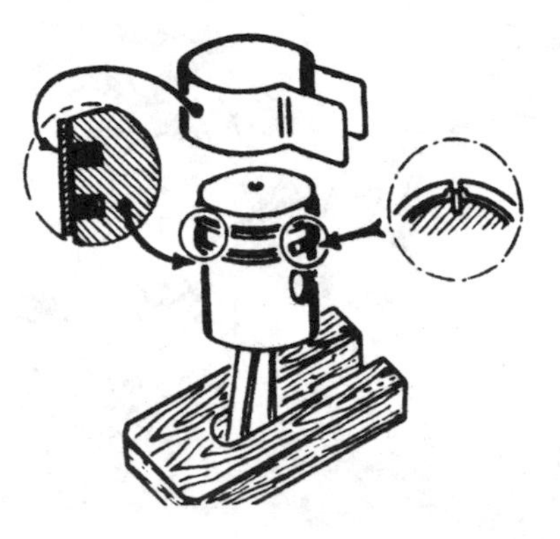

FIGURE 7-33. *A homemade ring clamp and a wooden block make ring installation easier for engines with detachable cylinder barrels.*

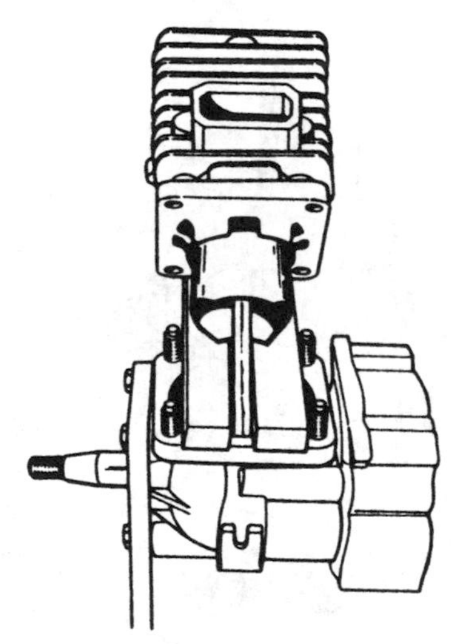

FIGURE 7-34. *A compressor is not needed if the bore has a taper on the lower edge.*

Cylinder bores

All discount-house and a handful of upper-echelon engines, such as the Briggs 11-CID Intec, run their pistons directly against the aluminum block metal. While soft metal cylinders can be rebored to accept oversized pistons (chrome-plated to reduce scuffing), the exercise seems futile. Aluminum-bore engines have a working life of 200 hours or so. It is possible to upgrade these engines with cast-iron cylinder liners. Expect to pay $80 to $150 for this service. In the discussion that follows, I assume you are working with cast iron.

Examine the bore for deep scratches, aluminum splatter from piston melt, and for the cat's tongue texture that comes from silicon particles ingested past a faulty air cleaner. Maximum wear occurs just below the upper limit of ring travel, where heat is greatest, lubrication minimal, and corrosives most concentrated. In the past, upper cylinder wear could undercut the bore enough to leave a ridge. Thanks to modern lubricants, the ridge and its corrective, the ridge reamer, have pretty well passed into history.

It was once considered necessary to roughen the cylinder with a hone to seat new rings. Some manufacturers continue to insist on honing; others say that wear, however induced, is wear. The decision is up to the rebuilder.

Boring cylinders is a job that should be relegated to an automotive machinist who has the tools and set up expertise to bore at 90° to the crankshaft centerline.

Confusion arises because Briggs and a few other small-engine makers size replacement pistons relative to the diameter of the original piston. A Briggs piston stamped ".030" measures thirty thousandths of an inch larger than the standard piston. If the bore is machined 0.030-in. over its original diameter, the replacement piston will have the correct running clearance. Automotive practice is to base piston oversizes on the bore, which means that a piston marked ".030" is 0.030 in. larger than the original bore diameter. For the piston to have room for expansion, the bore must be machined to 0.032 or 0.033 in.

Connecting rods

Cast or forged aluminum is the material of choice for four-cycle rods. Utility and light-duty engines run their crankpins directly against rod metal. Figure 7-35 illustrates this type of construction. Some manufacturers supply undersized rods so that the crankpin can be reground.

Better quality rods have precision bearing inserts at the big end and a brass or bronze bushing at the small end (Fig. 7-36). Under-sized inserts (0.010 and 0.020 in. for American-made engines) permit the crank to be reground.

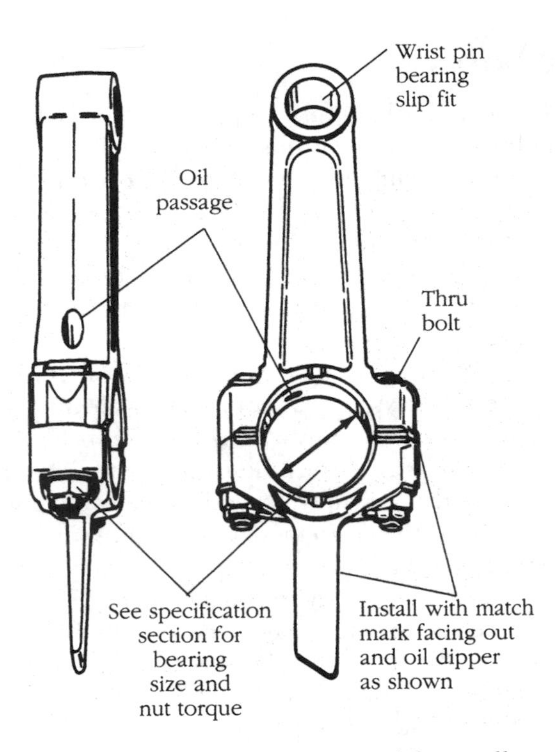

FIGURE 7-35. *Aluminum is a favorite material for small-engine rods.*

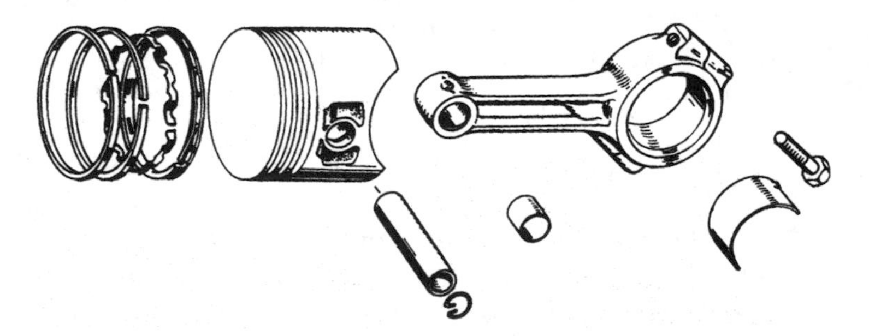

FIGURE 7-36. *Onan connecting rods feature a brass-bushed small end and precision inserts at the big end.*

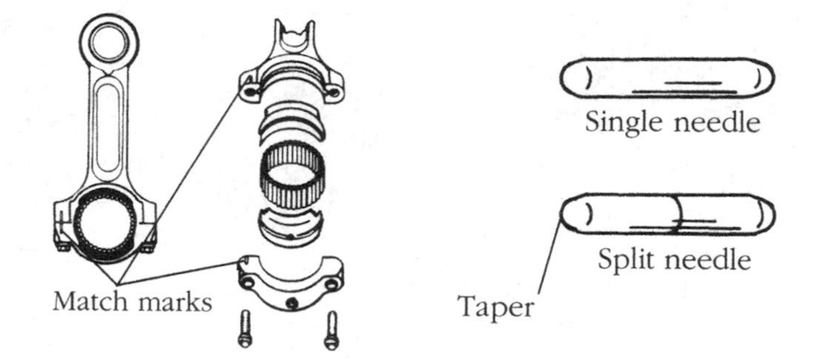

FIGURE 7-37. *Two-stroke rod with needle bearings at the big end and a brass bushing at the eye.*

Figure 7-37 shows an aluminum rod for a two-cycle engine with single or split needle bearings that run on steel races at the big end, and a bushing at the rod eye. String trimmers and the like use stamped-steel rods, which cost less than aluminum.

Catastrophic rod failure usually involves the big-end bearings. Plain bearings skate on a wedge of oil that develops soon after startup. Once up to speed, the bearing makes no contact with its journal. Insufficient bearing clearance prevents the oil wedge from forming; too much clearance causes the wedge to leak down faster than it can be replenished. In either case, the result is metal-to-metal contact, fusion, and a thrown rod.

Needle bearings make rolling contact against their races without the cushion of an oil wedge. Consequently, any discontinuity—fatigue flaking, a spot of rust, skid marks—results in bearing seizure.

Incorrect assembly can also result in rod breakage. Big ends crumble into bite-sized chucks when the fasteners have insufficient preload. Proper torque might not have been applied during assembly or the rod locks might have failed, allowing the bolts to vibrate loose. This is why manufacturer's torque specifications have the authority of Holy Writ and why new locks or lock nuts should be installed whenever a connecting rod is disassembled. Bend-over tab locks usually carry a spare tab that can be used during the first overhaul. Once a tab has been engaged, it should not be straightened and reused.

Rod orientation

Correct orientation has three components:

- Piston-to-rod. As mentioned earlier, the piston pin may be offset relative to the piston centerline. Wrong assembly results in knocking.

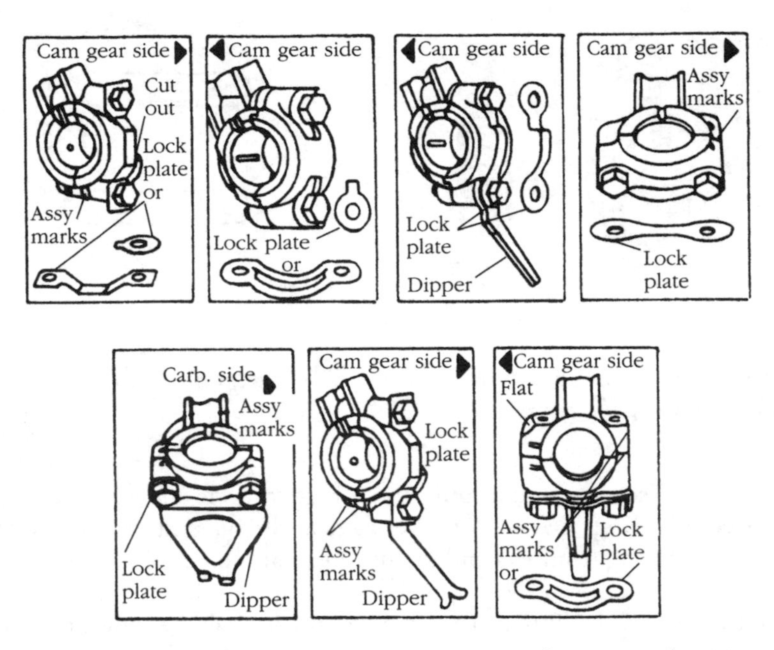

FIGURE 7-38. *Briggs & Stratton rod-to-engine and cap-to-rod orientation. McCulloch and a few other manufacturers fracture their rod caps after machining. When assembled correctly, the parting line becomes almost invisible.*

- Rod-to-engine. Some connecting rods are drilled for oil transfer; others are configured so that reverse installation locks the crankshaft.
- Cap-to-rod shank. In order to maintain precision, rods and caps are assembled at the factory and reamed or diamond-bored to size. Stamped or embossed marks identify cap orientation (Fig. 7-38). Failure to assemble the cap correctly results in early and catastrophic failure.

Rod inspection

The piston should pivot of its own weight on the rod eye when held at 45° off the vertical at room temperature. Pin-to-piston fits are tighter, but loosen when the piston reaches operating temperature. In no case should the piston wobble or tilt on its pin. Replacement rod-eye bushings sometimes install without the need for finish reaming, but do not count on it.

The big-end bearing is the most critical rubbing surface and never more so that when the bearing consists of needles or rollers. Any discontinuity on the crank pin means that both the crankshaft and rod bearings must be replaced if the engine is to live. A single rust pit sets in motion a chain of events that culminates in rod seizure.

Measure the crank pin at several places along its length and around its circumference with a good-quality micrometer—dial or electronic calipers do not have the requisite precision. Taper and out-of-round should be 0.001 in. or less. Do the same for plain-bearing connecting rods. The difference between rod ID and crankpin OD is the running clearance, which should be no more than 0.0030 in.

Caution: Do not attempt to restore bearing clearances by filing the rod cap.

That said, most mechanics approximate main-bearing clearances with plastic-gauge wire, available in various thicknesses from auto parts stores. Sealed Power SPG-1 Plastigage comes in three color-coded sizes—green reports a clearance range of 0.002 in. to 0.003 in., red spans 0.002 in. to 0.006 in., and blue 0.004 to 0.009 in. Everyone, even those who have access to precision measuring instruments, should use the wire as an assembly check.

Follow this procedure:

1. Turn the crankshaft to bring the rod to bottom dead center.
2. Remove the rod cap.
3. Wipe off any oil on the rod and crankpin.
4. Tear off a piece of green gauge wire and lay it along the full length of the journal (A in Fig. 7-39)
5. Install the rod cap, with match marks aligned, and torque down evenly to factory specs. The crankshaft must remain stationary as the bolts are tightened.
6. Remove the cap and compare the width of the wire against the scale printed on the envelope (B in Fig. 7-39). Average width corresponds to bearing clearance; variations in width indicate the amount of crankpin taper.
7. Repeat the process with two pieces of gauge wire across the journal (C in Fig. 7-39). The relative widths of the wires are a crosscheck on taper and say something about out-of-round.
8. Scrape off all traces of the wire from the bearing and journal.

Caution: As gauge wire ages, it hardens and becomes less accurate. Shelf life is said to be about six months.

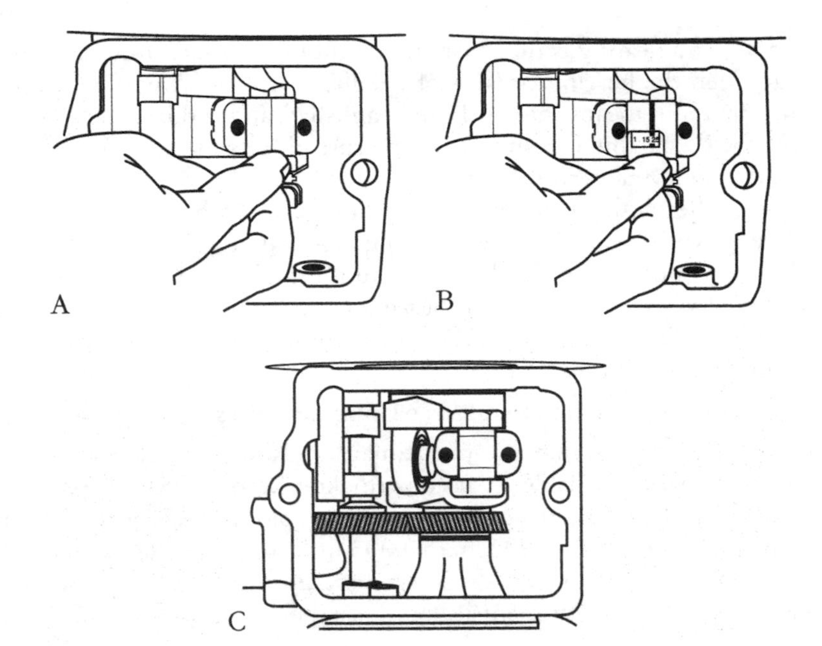

FIGURE 7-39. *Lay a piece of plastic gauge wire along the length of the crankpin (A). Assemble the cap and, without moving the crankshaft, torque the rod nuts to factory specs. Lift the cap off and measure the flattened wire against the scale on the package (B). Repeat the operation, positioning the wire at two points on the crankpin circumference to detect taper and out-of-round (C).*

Rod assembly

Coat all bearing surfaces with clean motor oil or assembly lube. Grease loose needle bearings to hold them in position around the periphery of the crankpin as the rod is installed (Fig. 7-40). All needles are accounted for if they pack closely around the crankpin with no space for another.

Check piston-to-block and piston-to-rod orientation one final time. Turn the crank to bottom dead center and guide the rod assembly home. Install the cap and verify its orientation. Tighten the rod bolts or studs evenly in three increments to specified torque.

Turn the engine over by hand to detect possible binds. The rod should move easily from side-to-side along the crankpin. Manufacturers do not often provide side-play specifications, but connecting rods need several thousandths of an inch of axial freedom.

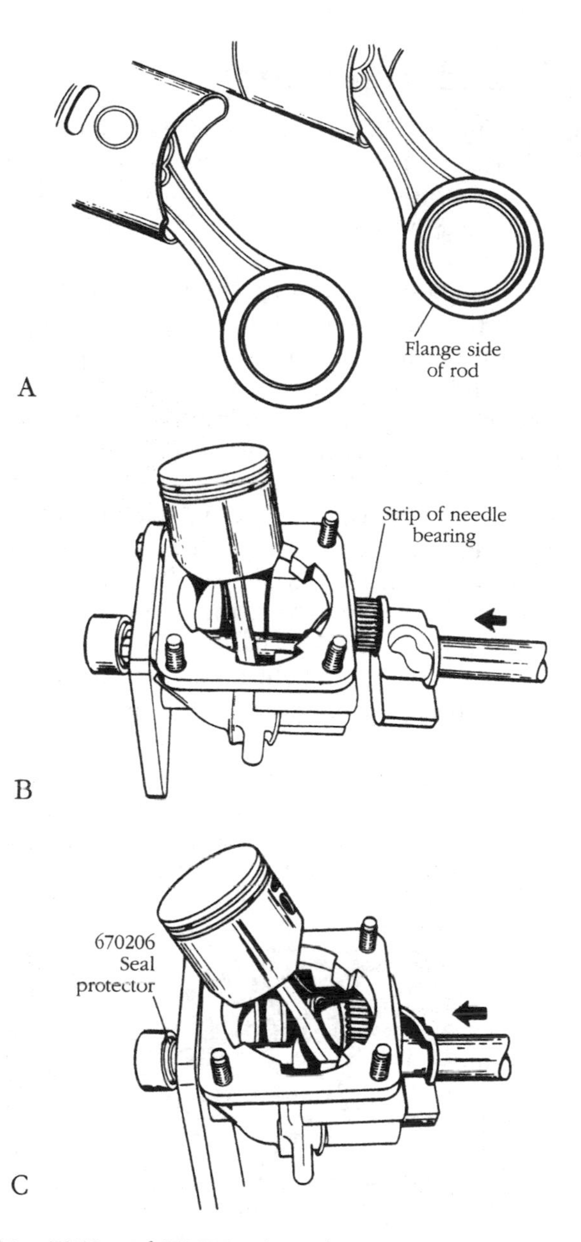

FIGURE 7-40. *TVS and TVXL840 rods present a special case. The rod installs with the flange toward the pto side of the engine (A). Grease-packed bearings go on the crankpin (B) and the rod is gingerly slipped over the crankpin and bearings (C). Note the factory-supplied seal protector, which is one of two needed for this job.*

Crankshafts and cam timing

It is always good practice to align timing marks before four-cycle engines are disassembled. For most engines, crankshaft and camshaft timing marks index at top dead center on the compression stroke. Secondary marks on rotating-balance and accessory-drive shafts index to the crank or cam after the valves are timed.

If the marks are missing or ambiguous, time from the “rock” position. Rotate the crankshaft to bring No. 1 piston to top dead center on what will become the compression stroke. Install the camshaft, slipping it under the tappets. Rock the crankshaft a degree or two on each side of tdc, alternately engaging the intake and exhaust valves. Timing is correct when the free play in crankshaft movement splits evenly between the two valves. If one valve leads the other, reposition the camshaft one tooth from that valve.

Ball- and roller-bearing cranks used on four-cycle industrial engines can present something of an extraction problem. Because of the confined quarters, the camshaft must be dropped out of position to maneuver the crankshaft out of the block. These engines drive the camshaft from the magneto side or hide the timing mark behind a ball bearing. Some manufacturers stamp a mark on the crankshaft web (which makes alignment problematic) and others bevel the associated crank-gear tooth (Fig. 7-41).

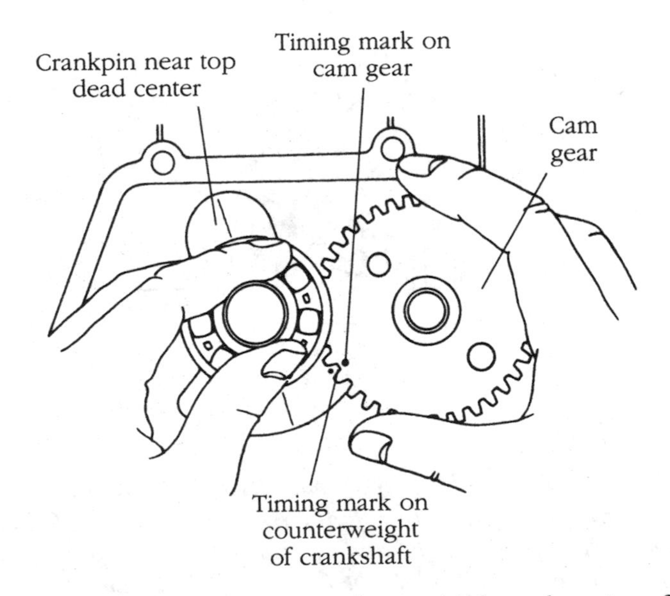

FIGURE 7-41. *Timing marks are not always visible at the point of tooth contact.* Briggs & Stratton Corp.

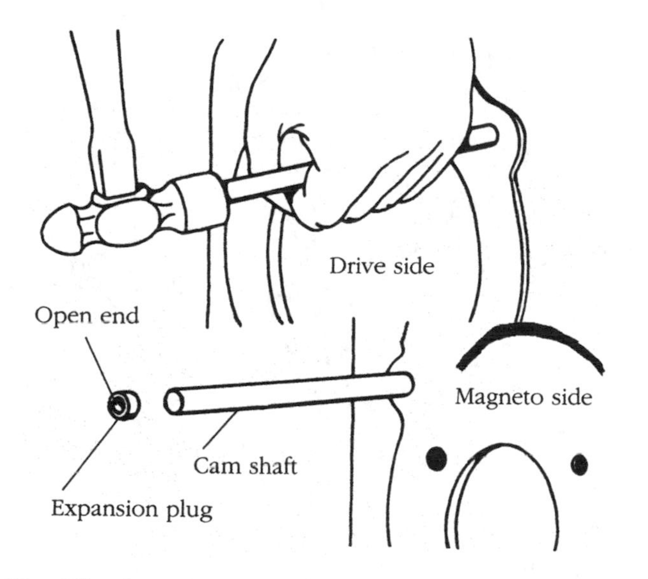

FIGURE 7-42. *The classic Briggs and Kohler camshaft axle drives out through an expansion plug, which should be coated with a sealant before the cam is installed.*

Release the camshaft by driving out the cam axle, as shown for a Briggs engine in Figure 7-42. Classic Kohlers follow the same pattern. Note the expansion plug is coated with sealant before assembly. Timing goes easier if you color the associated pair of crankshaft teeth with a Magic Marker.

Figure 7-43 illustrates crankshaft inspection points. Give special attention to the crankpin, as described earlier. When the crank is drilled for pressure lubrication, it is good practice to remove the expansion plugs and clean the oil passages, which act as sludge traps.

Lightly polish the journals with No. 600 wet-or-dry emery paper saturated in oil. To avoid creating flat spots, cut a strip of emery paper as wide as the journal. Wrap the strip around the journal, and spin it with shoelace or leather thong.

Straightening crankshafts is a touchy subject fraught with legal complications for the mechanic who gets someone hurt. Even so, experienced and patient craftsmen routinely straighten cranks bent a few thousandths. If you want to pursue this matter, recognize that you are on your own.

Warning: No small-engine manufacturer recommends that crankshafts be straightened.

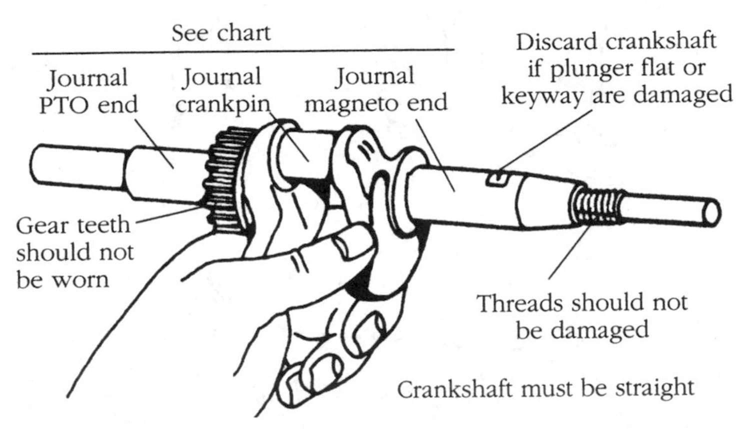

FIGURE 7-43. *Briggs crankshaft inspection procedure applies to other makes, with the proviso that some cranks are drilled for pressure lubrication. Clean the oil passages and lightly chamfer the oil ports.*

The work requires two machinist's V-blocks, a pair of dial indicators, and a straightening fixture usually built around a hydraulic ram. Tremendous forces are involved. The crank is supported on the blocks at the main bearing journals with the indicators positioned near the ends of the shaft. Total run out should be no more than +/−0.001 in. (0.002 in. indicated). Using the fixture, bring the crank into tolerance in small increments with frequent checks. Once the indicators agree, send the shaft out for magnetic-particle inspection. Skipping this final step, which only costs a few dollars at an automotive machine shop, can be disastrous for all concerned. Crankshafts break, especially when bent and straightened.

Upon assembly, check endplay, or float. This check is made internally (Fig. 7-44) with a feeler gauge or from outside the engine with a dial indicator. The amount of float is not crucial so long as the shaft has room to expand. Specs fall in the 0.004 to 0.009 in. range. A doubled up or thicker flange/side-cover gasket increases float when the dimension has been reduced by a new crank or flange casting. A thrust washer—usually placed between the crank and pto main bearing and, occasionally, on the magneto side—compensates for wear.

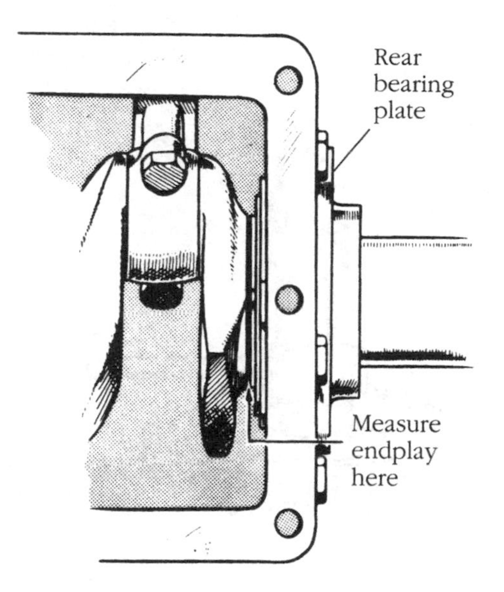

FIGURE 7-44. *Crankshaft float measured with a feeler gauge on an Onan engine. A suitably mounted dial indicator may also be used.*

Camshafts

The camshaft lives in the block on side- and overhead-valve engines. As shown back in Figure 7-42, the cams for Kohler Magnum and vintage Briggs engines ride on a steel pin. Other engines run their cams on plain bearings in the block and cover.

Cam failure of the kind that most mechanics flag is obvious: once the surface hardness goes on iron cams, the lobes rapidly wear round. Gear teeth occasionally fatigue and break. Interestingly enough, Briggs reports that its plastic cams generate fewer warranty claims than the metal versions.

Many camshafts include a compression release to aid starting (Fig. 7-45). These systems employ a pin or other protrusion that lifts one of the valves during cranking. When the "bumper" wears, the factory fix is to replace the camshaft. However, a welder can usually build up the worn surface with hard facing.

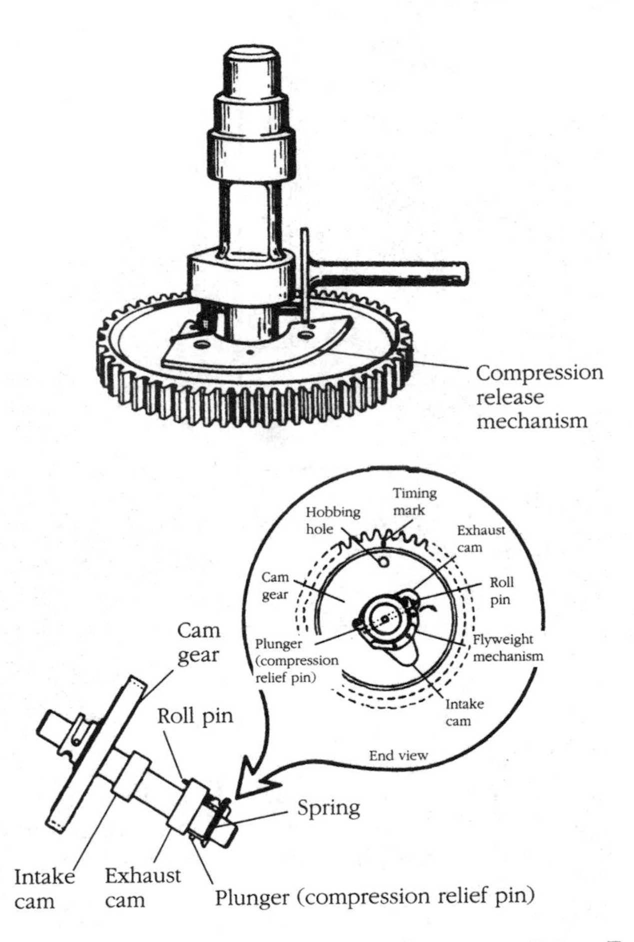

FIGURE 7-45. *Compression releases come in three types. Briggs Eazy-Spin employs a ramp ground on the cam profile that unseats the intake valve. These units give no problems. Others employ a bumper, either spring-loaded as shown or actuated by a linkage from the starter, to unseat one of the valves during cranking. The bumper is the weak spot.* Tecumseh Products Co.

Main bearings

The crankshaft runs on plain or anti-friction (ball or roller) bearings, or a combination of both types.

Antifriction bearings

Figure 7-46 shows a typical setup using two tapered roller bearings with provision for a hardened washer at the magneto side to control endplay. Check the condition of the bearings by removing all traces of lubricant and spinning the outer races by hand. Roughness or a tumbler-like noise means that the bearings have reached the end of their service lives.

Caution: Do not spin anti-friction bearings with compressed air.

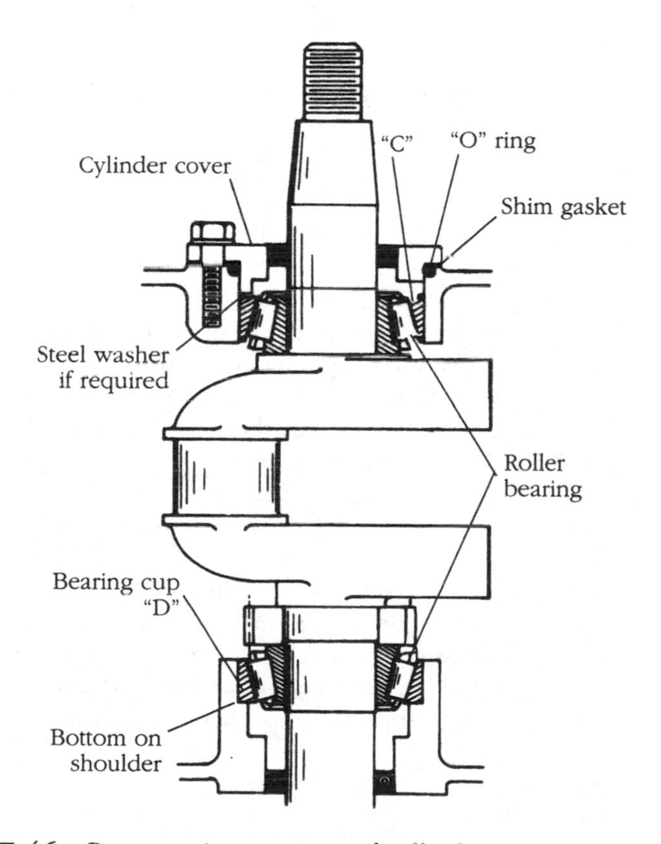

FIGURE 7-46. *Better engines use tapered roller bearings (as opposed to balls) that absorb large amounts of thrust as well as radial loads.*

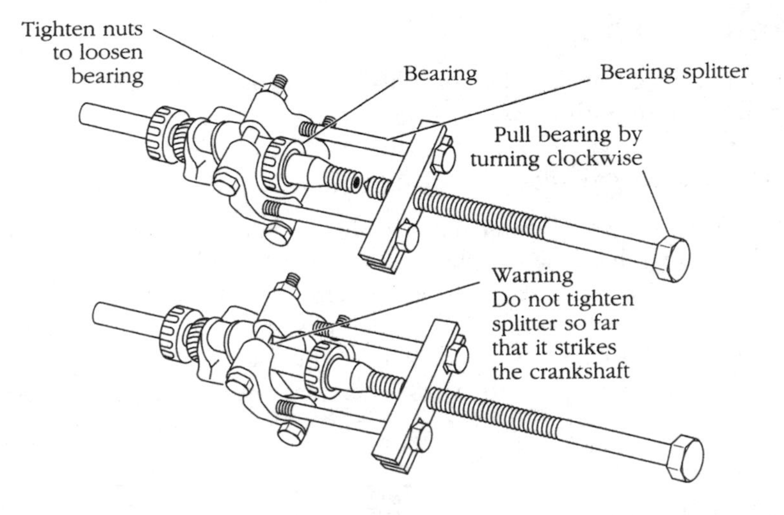

FIGURE 7-47. *Anti-friction bearings remain on the crankshaft unless they will be replaced.*

Extract defective bearings with a splitter (Fig. 7-47). Once drawn in this manner, the bearings cannot be reused. The preferred method of installation is to heat bearings in a container of oil until the oil begins to smoke (corresponding to a temperature of about 475° F). A wire mesh supports the bearing off the bottom of the container.

The more usual technique is to press the bearing cold while supporting the crankshaft at the web and applying force to the inner race. Figure 7-48 illustrates this operation for Kohler double-press bearings. These bearings are first pressed into their covers with the arbor on the outer race and then over the crank with the arbor on the inner race.

Anti-friction bearings seat flush against the shoulders provided. Upon assembly, check endplay against specification and adjust as necessary with shims or gaskets.

Anti-friction bearings can be purchased from bearing supply houses at some savings over dealer prices. Be certain that the replacement matches the original in all respects. Unless you have reliable information to the contrary, do not specify the standard C1 clearance for bearings with inner races. Ask for the looser C3 or C4 fit to allow room for thermal expansion.

Plain bearings

In a perfect world, one would establish main-bearing clearances with inside and outside micrometers as described for crankpin bearings. That said, I have yet to see a small-engine mechanic do more than wobble the crankshaft. Most plain bearings are set up with 0.002-in. clearance new and tolerate something like twice that before the seals wear out.

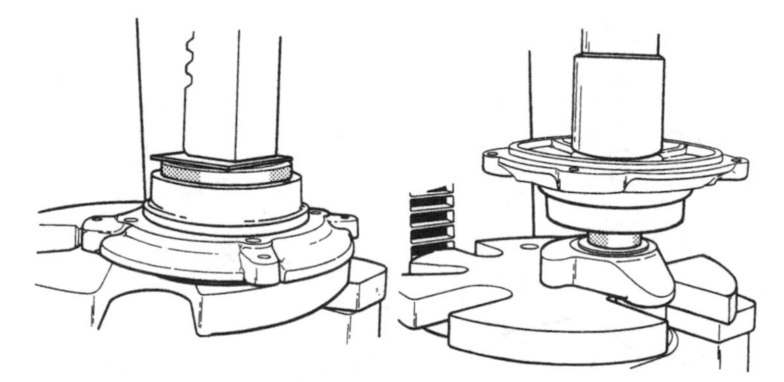

FIGURE 7-48. *Some Kohler engines use pto-side bearings with a double interference fit. The cup, or outer race, presses into the bearing cover and the inner race, together with the bearing and cover, presses over the crankshaft. A support under the crankshaft web nearest to the arbor isolates the crankpin from bending loads.*

Engines from major manufacturers can be rebushed, but the work is best left to a dealer who has the proper reamers and pilots. Shops that cater to racers can install Briggs DU™ Teflon-impregnated bronze bushings that withstand twice the radial loads of aluminum-block metal bearings.

Thrust bearings

Most manufacturers install a hardened steel washer between the flange/side cover and crankshaft cheek. Kohler and a few others specify proper babbit or roller thrust bearings. Poorly maintained vertical-shaft engines develop severe galling at the flange thrust face, which can be corrected by resurfacing or replacing the casting.

Seals

Seals, mounted outboard of the main bearings, contain the oil supply on four-cycle engines and hold crankcase pressure on two-strokes. Seal failure is signaled by oil leaks at the crankshaft exit points or, on two-cycle engines, by hard starting and chronically lean mixtures.

The old seals pry out with a flat-bladed screwdriver (Fig. 7-49). Install the replacement with the maker's mark visible and the steep side of the elastomer lip toward the pressure. Lubricate the lip with grease. If you coat the seal OD with sealant, be careful not to allow the sealant to contaminate the seal lips or clog the oil-return port.

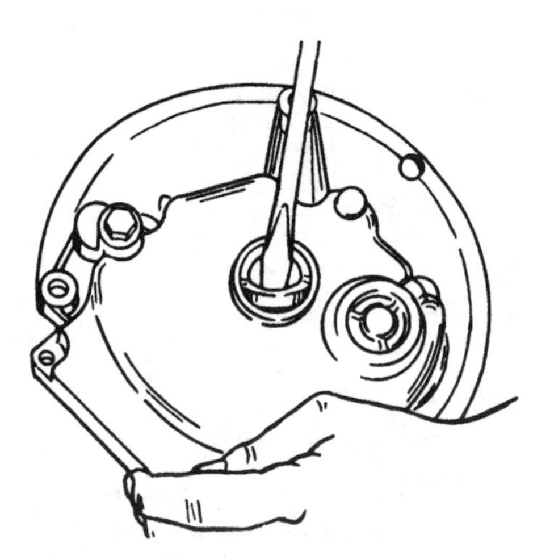

FIGURE 7-49. *Crankshaft seals come free with the help of a large screwdriver.* Tecumseh Products Co.

Installation is best done with a driver sized to match the OD of the rim (Fig. 7-50), although a piece of 2 × 4 works in an emergency. Drive the seal to the original depth (usually flush or slightly under-flush) unless the crankshaft exhibits wear from seal contact. In that case, adjust the seal depth to engage an unworn area on the crank, but do not block the oil port in the process.

The crankshaft must be taped during installation to protect seal lips from burrs, keyway edges, and threads. Cellophane tape, because it is thin, works best.

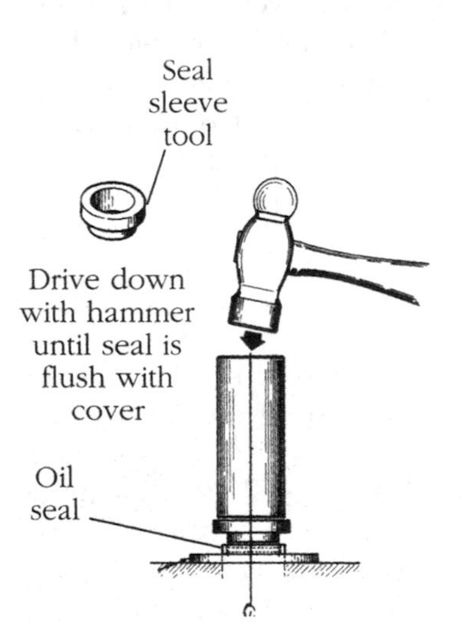

FIGURE 7-50. *The correctly sized driver confines installation stresses to the outer edge of the seal retainer.*

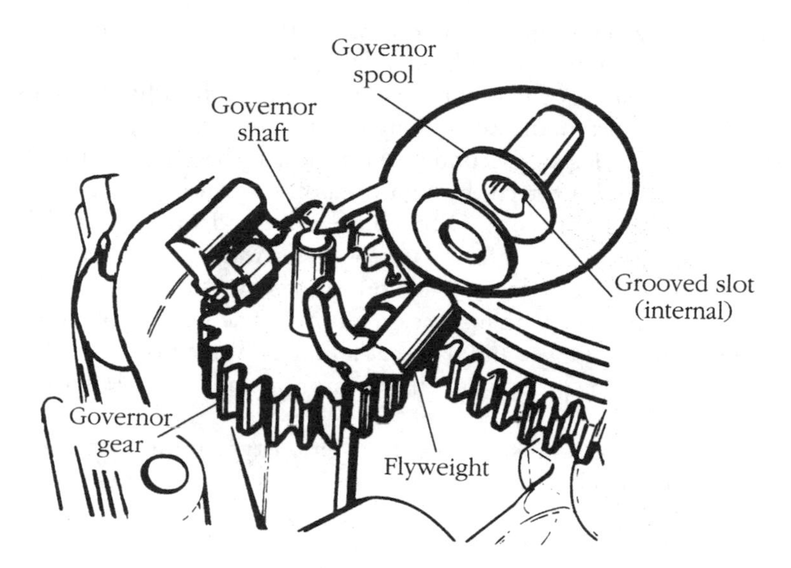

FIGURE 7-51. *Typical centrifugal governor. Flyweights react against a plastic spool.*

Governor

The unit shown in Figure 7-51 is typical of most centrifugal governor mechanisms. Paired flyweights, driven at some multiple of engine speed, pivot outward with increasing force as rpm increases. The spool translates this motion into vertical movement that appears as a restoring force on the carburetor throttle linkage.

Work the mechanism by hand, checking for ease of operation and obvious wear. The governor shaft presses into the block or flange casting; should it need replacement, secure the shaft with Loctite bearing mount and press it to the original height.

Lubrication systems

Lubrication systems require careful scrutiny. Conscientious mechanics will not release an engine unless all circuits have been traced, cleaned with rifle brushes, and buttoned up with new expansion plugs.

Any of three oiling systems are used. Most side-valve engines employ a dipper on the end of the connecting rod or a rotating slinger to splash oil about in the crankcase.

Semi-pressurized systems combine splash with positive feed to some bearings and, when present, to overhead-valve gear. Figure 7-52 illustrates the Tecumseh approach. A small plunger-type pump, driven by the camshaft, draws oil from a port on the cam during the intake stroke (Fig. 7-53). As the plunger telescopes closed, a second port on the camshaft hub aligns with the pump barrel and oil is forced through the hollow camshaft to a passage on the magneto side of the block. Oil then flows around a pressure relief valve (set to open at 7 psi) and into the upper main-bearing well. Most models have the crankshaft drilled to provide oil to the crankpin.

Blow out the passages and inspect the pump for scores and obvious wear. Replace the pump plunger and barrel as a matched assembly.

Caution: Prime the pump with clean motor oil and assemble with the flat side out.

Other semi-pressurized systems use an Eaton-type pump, recognized by its star-shaped impeller. The pump cover usually shows the most severe wear.

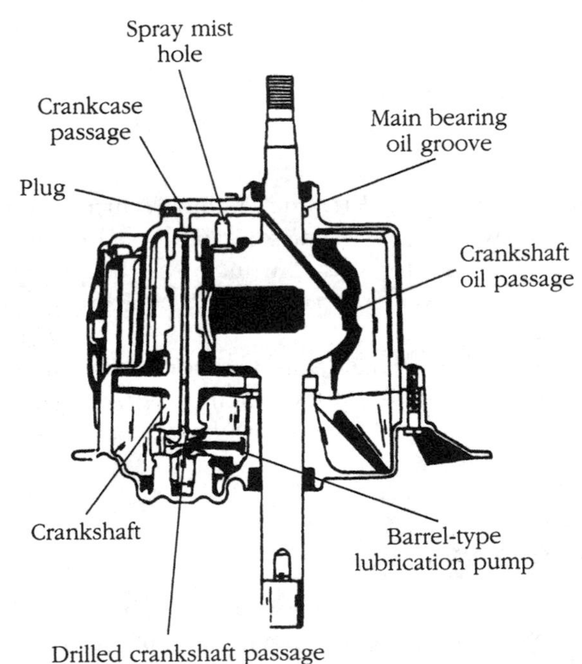

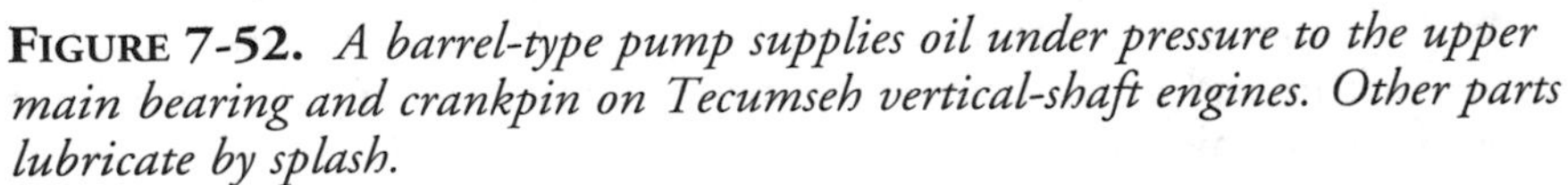
FIGURE 7-52. *A barrel-type pump supplies oil under pressure to the upper main bearing and crankpin on Tecumseh vertical-shaft engines. Other parts lubricate by splash.*

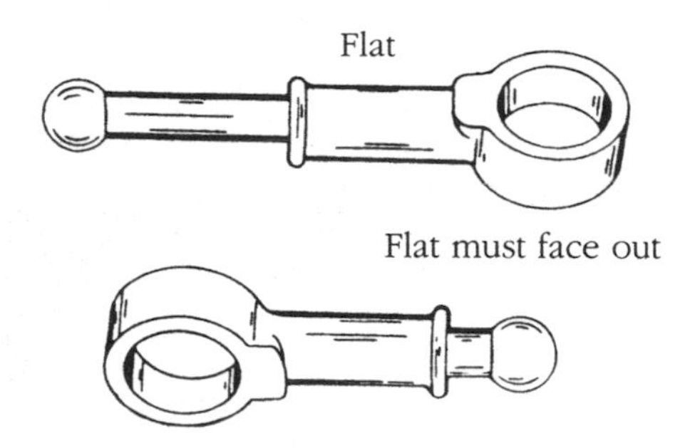

FIGURE 7-53. *Plunger pump drives off a camshaft eccentric.*

Full-pressure systems deliver pressurized oil to all crucial bearing surfaces, although some parts receive lubrication from oil thrown off the crankpin and by oil flowing back to the sump. The Kohler system is typical (Fig. 7-54). A gear-type pump supplies oil to the pto-side main bearing, crankpin, and camshaft. The hollow camshaft carries oil to the magneto-side main bearing and crankpin. A pressure-relief valve limits pressure to 50 psi.

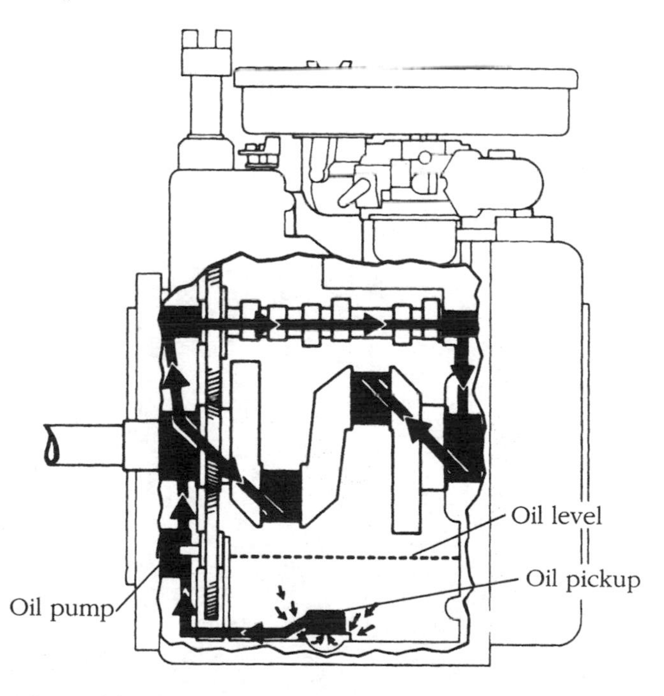

FIGURE 7-54. *Kohler full-pressure system utilizes a gear-driven pump and a hollow camshaft.*

Appendix Internet resources

Information about small engines on the Internet varies in quality. Most of it comes down to a form of advertising. But once you winnow out the chaff, the web becomes an unparalled source of information. How did we ever live without it?

Sites listed below are ones that I have found useful and generally reliable, although the quality of forum discussions varies with the participants.

Engine manufacturers

Arrow

Arrow at http://www.arrowengine.com manufactures single- and twin-cylinder cast-iron block engines of legendary durability. Some C series engines have been in operation in the oilfields for 75 years. It's not unusual for an Arrow to run for decades with no more than oil and spark plug changes. The website provides free service, operator's and parts manuals in .pdf format.

Briggs & Stratton

The corporate site, http://www.briggsandstratton.com/maint_repair/manual_and_more/, includes downloadable owner's manuals, illustrated parts lists (with part numbers and prices), and generator wiring diagrams. Hard copies of Briggs service manuals are available for purchase. Serious mechanics might want a copy of *Major Engine Failure Analysis,* which is well worth the $14.95 Briggs asks for it. The company also markets CD-ROM training aids.

Honda

Honda Europe, http://www.honda-engines-eu.com/en/welcome.html, provides down loadable owner's manuals and excerpts of service manuals, which unfortunately do not go very deeply into the subject.

Honda dealers no longer inventory service manuals. (Factory mechanics use CD-ROMs.) Printed versions of these manuals can be purchased from Helm, Inc. http://www.helminc.com/helm/homepage.asp?r

Kawasaki

An illustrated parts list can be found at the Kawasaki factory site, http://www.buykawpower.com, or at http://www.smallenginesuppliers.com. The Outdoor Power Equipment site http://www.johnfvining.com lists make-up torques, spark-plug gaps and types, valve clearances, and minimum compression readings for Kawasaki and other engines.

Kohler

http://www.kohlerengines.com offers free online shop and owner's manuals, and hard copies for purchase. The Kohler site also features a parts lookup, accessible with Microsoft Internet Explorer 5.5 and later browsers.

Onan

http://www.cumminsonan.com/engines/ deals with multi-cylinder Onan engines used to power Lincoln and Miller welding machines. Parts and service manuals for these engines must be purchased through a Cummings Onan distributor.

http://www.ssbtractor.com/ markets hard copies of service manuals for Onan engines fitted to small and not-so-small tractors. A downloadable service manual for 16, 18, 20 and 24 Hp CTM2 can be purchased at http://www.ecrapusa.com/.

Subaru-Robin

http://www.robinamerica.com/ provides downloadable operators, service, and parts manuals for all engine models at no charge.

Tecumseh

The corporate website http://www.tecumsehpower.com/ has a downloadable troubleshooting manual for the company's engines and Peerless hydrostatic transmissions. The factory store sells hard copies of service manuals and high-demand replacement parts.

Downloadable Tecumseh service manuals can be purchased from http://www.outdoordistributors.com/ or from http://www.hobbytalk.com and http://www.ecrapusa.com/index.php?main_page=conditions

Wisconsin

http://www.wisconsinmotors.com/ has parts lists for engines dating from the 1980s. Shop manuals are available for purchase at http://www.ssbtractor.com/ and for vintage models at the at American Small Engines Collector's Club http://www.asecc.com/

Bargains

Mower, tractor and other original equipment manufactures cannot judge demand for their products with precision. At the end of the season, surplus engines end in the hands of liquidators for sale on Internet. Tulsa Engine Warehouse, http://tewarehouse.com, boasts of a multi-million dollar inventory. Prices are significantly lower than dealer list. For example, the Small Engine Warehouse, http://www.smallenginewarehouse.com, lists 6-hp ohv Briggs Intecs for $225 with free shipping.

Forums

Do It Yourself, http://forum.doityourself.com/forumdisplay.php?f=70 focuses on outdoor power equipment.

Hobby Talk, http://www.hobbytalk.com, hosts forums on hand-held two-strokes (weed eaters, trimmers, etc.), which rarely get the attention they deserve. The site also has four-cycle and swap forums.

Nabble/MC Engine Design, http://www.nabble.com/MC-Engine-Design-f13886.html, serves as a roundtable for experts who debate ways of extracting more power from motorcycle engines. It's a privilege to listen in.

At last count, the PER Small Engine Forum contains more than 6600 topics http://www.perr.com/forum/viewforum.php?f=2&sid=e668b4f61263daf285215 3bdad309d3b.

Small Engine Talk, http://smallenginetalk.com/, includes brand-specific forums for all popular makes, including Kawasaki for which not much information is available.

Smokstak, http://www.smokstak.com/forum/, hosts several forums for vintage engines. This is the place to go if your Maytag magneto gives problems.

Tractors Forum, http://www.ssbtractor.com/, focuses on small tractors and their power plants. As indicated previously, the site provides service manuals for many current and obsolete engines.

Repair information

PER Notebook http://www.perr.com/tip.html contains 25 articles on air-cooled engines and generator service. Other repair information can be found at the Outdoor Power Equipment site http://www.johnfvining.com.

Carburetors

Aero-Cors-USA, http://www.aerocorsair.com/id27.htm, goes into detail about modifications to Walbro and Tillotson carburetors for paragliding. The same procedures apply to earth-bound applications.

The service manual for Tillotson's HS carburetor, widely used on hand-held tools, can be downloaded at http://www.tillotson-fuelsystems.com/manuals/hsmanual_us.doc.

Walbro's comprehensive service manuals are free for the downloading at http://wem.walbro.com/distributors/servicemanuals/.

USA Zama provides a comprehensive service manual, together with an applications guide at http://www.zamacarb.com/tips.html.

Ignition systems

Phelon Engine Electronics, http://www.phelon.com, manufacturers magnetos and high-tech ignition modules, several of which are microprocessor controlled. While these systems are not available for individual sale, the technology gives us a glimpse of what the future holds.

Miller's Small Engine & Speciality Shop, http://hometown.aol.com/pullingtractor/a1elect.htm, supplies ignition updates for battery and magneto systems.

http://www.jetav8r.com/Vision/Ignition/CDI.html is an excellent article, beautifully illustrated, on ignition theory

A three-part series on the CAFE Foundation site, http://www.cafefoundation.org/, examines ignition as a means of reducing aircraft fuel consumption. The foundation, which is privately supported, asks that you make a contribution before downloading the material.

Vintage engines

The Antique Small Engine Collectors Club, http://www.asecc.com, is a large site with specifications for early Briggs and Johnson Iron Horse engines, an active forum, and hundreds of photos. ASECC also markets repair manuals for vintage Briggs, Kohler, Tecumseh, Continental and Wisconsin.

Doug's Reo Engine Site, http://www.geocities.com/reo_engine/, reproduces pages from the Reo service manual.

Erv's Reo Engine Site, http://members.aol.com/reo43/, includes a history of Reo and a list of models.

John Cox's Site, http://home.cogeco.ca/~jcox109/, is rich with information about the excitement of the hunt for vintage Briggs engines.

Leonard's Toys, http://www.oldengine.org/members/keifer/, features riding mowers, washing machine engines, and a rare sickle mower.

The Maytag Shed, http://www.maytagshed.com/, has photos and parts breakdowns of Maytag washing machine engines.

The Washing Machine Museum, http://www.oldewash.com/, must be the world's largest collection of vintage washing machines, many of which are gasoline-powered.

Probably the most remarkable vintage site is not about collecting or restoration. Flashback Fabrications, http://flashbackfab.com/index.html, tells the story of how Paul Brodie replicated the 1919 Excelsior Cyclone racing motorcycle. More than a hundred photographs record how the project evolved from initial Auto-Cad drawings to preparations for a record run at Bonneville.

Index

D

E

F

V

W